Lecture Notes in Earth Sciences

Edited by Somdev Bhattacharji, Gerald M. Friedman, Horst J. Neugebauer and Adolf Seilacher

35

Lars Håkanson

Ecometric and Dynamic Modelling –

Exemplified by Caesium in Lakes After Chernobyl

Methodological Aspects of Establishing Representative and Compatible Lake Data, Models and Load Diagrammes.

Springer-Verlag

Berlin Heidelberg New York London Paris
Tokyo Hong Kong Barcelona Budapest

Author

Prof. Lars Håkanson
Dept. of Hydrology, Uppsala University
V. Ågatan 24, S-752 20 Uppsala, Sweden

ISBN 3-540-53997-2 Springer-Verlag Berlin Heidelberg New York
ISBN 0-387-53997-2 Springer-Verlag New York Berlin Heidelberg

Printing and binding: Druckhaus Beltz, Hemsbach/Bergstr.
2132/3140-543210 – Printed on acid-free paper

This publication is based on a report in Swedish by L. Håkanson, H. Kvarnäs, T. Andersson, G. Neumann and M. Notter. That report is based on a compendium produced for a course for the colleagues of the Liming-Mercury-Caesium project. The aim of the course was to try to establish a "common language, framework and sense of direction" for the work we were doing. This book

is dedicated to

the colleagues in the Liming-Mercury-Caesium project

because the ultimate basis for this publication is their excellent work.

TABLE OF CONTENTS

1. INTRODUCTION

Ecometry concerns measurements and interpretation of ecological data and relationships between data. It deals with most matters involved in the scientific aspects of the representativity and information value of samples and does not, in fact, concern statistical methods. In particular, ecometry can be regarded as an approach to obtain so-called load models and **load diagrammes** (effect-dose-sensitivity diagrammes), which are one of the aims/final products in aquatic environmental consequence analysis (Håkanson, 1990; all these terms will be explained later on). This publication is meant to demonstrate what can and **cannot** be done using ecometric approaches. It must be emphasized at the outset that the main intention here is **not** to provide new radioecological knowledge on how Cs-137 is dispersed in aquatic ecosystems after the Chernobyl accident and is taken up in fish, but to use Cs-137 as a type substance and pike as a biological indicator to go through methods which should also apply to other types of environmentally hazardous substances (it could just as well have been substance X in ecosystem Y). As a secondary effect, we may also learn something about Cs-137. Several terms and methods, which have not been used earlier in the aquatic environmental sciences, e.g., ecometric analysis and dynamic modelling using moderators, will be discussed and defined.

The presentation will follow a path where the following main questions are asked:

⇨ How can a load diagramme be developed for caesium in pike?

⇨ How can models be developed to describe and predict the development of caesium in pike more generally? In this context, we will focus on ecometric modelling and on a type of dynamic modelling designed to describe both conditions in individual lakes and differences between lakes. These results should be taken for what they are — a first attempt. The models used are based, and should be based, on data that are normally available to environmental authorities and which can be used to describe and predict differences between lakes. An important question is: When will pike in the most contaminated lakes attain Cs-concentrations below the Swedish guideline of 1500 Bq/kg wet weight?

⇨ How can results from the ecometric analysis and the dynamic modelling be used to obtain a country-wide perspective (in the case of Sweden) on the Cs-concentrations in pike and perch?

Thus, the objective here is to **not to make a survey of the literature concerning radioactive substances in lakes**.

Another aim of this booklet is to discuss the reliability of empirical water quality data. This is important for applied work, e.g., monitoring programmes, as well as theoretical research, e.g., validating models. The principles are discussed using an extensive set of data from Swedish lakes on radioactive caesium in pike, perch, water and sediments, and a broad set of limnological data (pH, Secchi depth, temperature, alkalinity, total-P, conductivity, Fe, Ca, hardness, chlorophyll-a and colour). These parameters generally vary in a lake, both temporally and areally. The focus here is on such variations within lakes and between lakes and how to express lake-typical values.

The results presented here emanate from a project where the following question was posed: Is it possible to obtain scientifically acceptable and practically applicable measures to speed up the natural decrease of radioactive caesium in lake fish after the Chernobyl accident of April 1986? Many factors may influence uptake and concentration of caesium in fish. In order to be able to predict that a certain remedy, e.g., liming, potash treatment or intensive fishing, has a certain effect on the Cs-concentration in fish, all these factors of potential influence must be monitored. This can only be achieved through comprehensive field studies, data processing and modelling according to a strictly structured approach. This publication presents the fundamentals for an ecometric approach which has been developed by and is used within the Swedish Liming-Mercury-Caesium project (Håkanson, 1990; Håkanson et al., 1990a,b).

As a background, it may also be said that, in 1987, more than 14000 lakes in Sweden had fish with Cs-concentrations above the level of 1500 Bq (=bequerel) per kg wet weight used by the National Swedish Food Administration as a retailing guideline (Andersson et al., 1990). The problem with radioactive caesium after Chernobyl has attracted wide interest in the mass media, not least in the local press, and in the international scientific literature (see Petersen et al., 1986; Persson et al., 1987; Santchi et al., 1990).

Ecometric models are meant to be based on solid scientific knowledge (on all relevant matters from methodological aspects of sampling and analysis to physical, geographical and ecological processes), reliable empirical data and statistical methods. These models are primarily used to quantify differences **between** entire lakes from simple, readily accessible standard parameters.

Dynamic models, on the other hand, are not based on statistical relationships but on a causal analysis. Such models can be constructed without empirical data. If dynamic models are to be used in practice, e.g., to quantify fluxes of energy, carbon or contaminants in lakes, the rates that govern the transport of the parameters between the various compartments in the lake ecosystem have to be known. This requires model calibrations against empirical data or laboratory data. In dynamic modelling, the analysis of dimension (of each parameter) is very important. Dynamic models are mostly used to study complex interactions and variations with time **within** defined ecosystems. The modelling requires experts in systems analysis/mathematics. Dynamic models are often difficult to validate and they sometimes tend to be develop "elephantiasis", i.e., grow to something very large.

Thus, ecometric models and dynamic models are used for different purposes; they do not compete, they complement each other. There are drawbacks and benefits with both types of model approaches. The presuppositions ("traffic rules") of the models must always be clearly stated. This publication is almost entirely focussed on ecometric models. Chapter 5, on dynamic models, has been included to demonstrate that there are ways to link ecometric and dynamic models. This link is created by means of a so-called moderator. Via the moderator, it is possible to construct models that combine some of the advantages of both dynamic and ecometric models.

During the work to prepare this book, several alternatives to demonstrate the various steps in the ecometric analysis were tested. From the very interesting topic of aquatic eutrophication, there are a lot of data available to illustrate, e.g., the concept of time compatible data (see Vollenweider, 1968; OECD, 1982; Rosenberg, 1986; Wallin et al., 1990). But had those data been used, so much explanation to the presuppositions would have to be included that the pedagogical benefits would be lost in a maze of circumstantial information. Data are also available on stable metals in lakes (see Håkanson et al., 1990b) to illustrate, e.g., how various mathematical transformations influence the relationships between parameters. But, once again, had these examples been used, so much circumstantial information would have had to be included that the pedagogical advantages would be lost. So, since a broad set of data on Cs-137 and other lake parameters is at my disposal, the idea here has been to use this coherent data-set to illustrate **all** the steps in the ecometric analysis.

Many of these steps are, really, rather simple. A hope with this publication is that the ecometric entity should amount to more than the sum of the parts. First, we will follow a main direction that leads to a load diagramme (Chapter 3). In Chapter 4 several exercises will be made where the ecometric approach is used in contexts where such approaches are not really meant to be utilized, namely to study variability with time. The intension is that these exercises will demonstrate the **limitations** with ecometric models. At the outset of each exercise, the set-up may seem reasonable, but at the end it may be shown that the approach gave unrealistic results.

It should also be stressed that this booklet is **not** meant to be either a textbook in statistics (there are already many such books at hand, see, e.g., Pfaffenberger and Patterson, 1987 or Neter et al., 1988), or a textbook in systems analysis or dynamic modelling (since there are also many excellent textbooks available in this sphere, see Vemuri, 1978 or Gustafsson et al., 1982). There is, in fact, only one statistical formula in this publication. **Thus, it is assumed that the reader is acquainted with basic statistical methods.**

The word *course* has two meanings: one is linked to students and teaching, the other to direction and stearing towards a goal. The publication is meant to be a course book in both these senses. That is, **a textbook** for graduate students of different "environmental" disciplines, for qualified technical personnel at environmental agencies and for scientists interested in ecosystems analysis from a multi-disciplinary perspective, as well as **a guidebook** on how to establish representative empirical data and functional groups/clusters, to distinguish between statistical and causal relationships, to quantify variations within and between ecosystems, to derive load models and load diagrammes and how to obtain a geographical perspective from a limited amount of data.

As a background, the reader will first (in Chapter 2) be acquainted with some basic concepts, the lakes and the data in a brief "introductory" chapter.

2. LAKES, DATA AND METHODS

2.1. EFFECT, DOSE AND SENSITIVITY

The terms effect, dose (=load) and sensitivity are meant to be general terms that may be applied to most contaminants in most environments, e.g., for nutrients, metals and organics in lakes, coastal waters and terrestrial ecosystems. The approach discussed here is based on an ecosystem perspective (entire lakes); there are other supplementary approaches, which range from tests at the cell level to computer models of what might happen in thousands of years to come (see, e.g., Mackey and Peterson, 1982; O´Neill et al., 1982; Cairns and Pratt, 1987; Landner, 1989).

The **effect term** is a key concept in this system. It is of vital importance that the reader realizes what is meant and not meant by the effect term. Different types of contaminants require different types of ecological effect parameters. The interest concerns: (1) which parts of the ecosystem are first damaged and (2) the concentrations at which the damage occurs. It is important to identify the weakest link in the ecochain and the concentration at which this link breaks, i.e., to identify the target indicators and the critical concentrations. The importance of identifying target indicators is illustrated in Fig. 2.1, which schematically shows that biological/ecological effects of the metal cadmium can be obtained in a very wide range of concentrations in short- and long-term laboratory experiments with various biological indicator organisms. The most sensitive indicator organisms react to Cd-concentrations in water of about 0.1-1 ppm, whereas the most Cd-resistant indicator organisms do not react until the concentration of Cd in water is 4-5 powers of ten higher.

For Cs-137 in lakes there are no generally accepted ecological/biological effect parameters. The threat does not appear to be directed against life in the lakes but is mainly directed at humans eating fish and then primarily at the foetuses of pregnant women and at small children. Adult men and women are not the primary target group. **This means that for Cs-137 one can focus on fish as indicator species.** The reason for focussing on perch and pike here is that it is generally rather easy to catch these species in Sweden. The fish are caught at several places in each lake. This gives a **lake-typical and not a site-typical value.**

In Sweden and in many other countries there is a traditional use of "1 kg pike" to blacklist lakes contaminated by mercury (see Lindqvist et al., 1991). Here we also focus on pike. **The Cs-concentration in pike is called an effect parameter**.

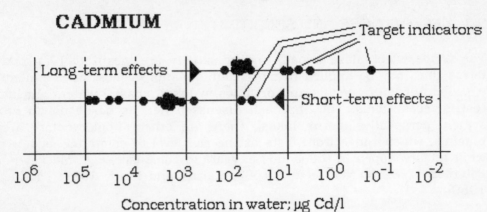

Test results for different biological organisms

CADMIUM

Fig. 2.1. Concentrations of cadmium yielding short- and long-term toxic effects on different biological indicator organisms (from SNV, 1980; and Håkanson and Jansson, 1983).

The term **dose** is used in different ways in different contexts, e.g., in radioecology, laboratory-based ecotoxicology and ecosystem-directed environmental science. Here, PRIMARY DOSE is considered to be the average fallout (in Bq/m^2) of caesium (see Fig. 2.2); whereas SECONDARY DOSE concerns the load of radioactive caesium in lakes from land runoff during a given space of time after the fallout event. It must be emphasized that the information on fallout in Fig. 2.2 may suffer from considerable errors for different types of soils (perhaps up to 2-3 fold, partly because different soils may cause considerable differences in measurement values).

The Cs-dose **to the lake** can be determined from water samples but the Cs-concentrations in water are low, the analysis is relatively expensive and many samples from different places and at different times must be made to get a representative lake-typical value. An alternative to water sampling is to use material collected in sediment traps (see Håkanson and Jansson, 1983).

The lake dose (=load=contamination) of Cs-137 has been estimated in several ways in this context:

1. By analysis of Cs-137 in samples of lake water. We have taken samples from central lake localities every and/or every second month during 1987 and 1988.

2. By analysis of the concentration of Cs-137 in material collected in sediment traps placed in central sites in the lakes. If the traps are placed in

Västernorrland
county

Gävleborg
county

*Fig. 2.2. Calculated deposition of caesium-137 (kBq/m²) over Sweden
and identification of the provinces of Gävleborg and Västernorrland (see
also Fig. 2.4). From Persson et al. (1987).*

the lake, e.g., 3 months, and the material deposited in the traps is
analysed for Cs, we obtain integrated **indirect** values concerning the dose
in the lake during the registration period. Here we report data from 1986,
1987 and 1988 from sediment traps placed close to the water surface (at
depths of 2 m) and close to the bottom (2 m above the bottom). These
data are presented as Cs-su87 or Cs-bo87 (in Bq/kg dry weight), where **su**
stands for "surface", **bo** for "bottom" and 87 for 1987.

The term **sensitivity** factors here refers to factors that may exert a direct or indirect influence mainly on the effect parameter, i.e., the lake mean concentrations of Cs in fish. It is important to emphasize that there are different types of relationships between different biological, chemical and geoscientific factors (see Fig. 2.3). Lake pH, e.g., directly and via known relationships, can be related to water chemical parameters such as alkalinity, amount of organic carbon, bioproduction, as well as to parameters describing the drainage area. This implies that, from a pure statistical viewpoint, it is difficult or impossible to determine "independent" limnological parameters - most parameters are related to each other in an extremely complex web of interactions. It is evident that it is very difficult to establish **causal relationships** in such complex ecosystems.

Here, we mostly look for sensitivity parameters that can be measured by standard analytical methods. Thus, target sensitivity parameters are those that both influence the spread and uptake of caesium in fish and those that can be modified by different practical remedial measures (like lake and wet land liming, lake fertilization, potash treatment and intensive fishing).

The following environmental factors will be discussed:

• Water chemical parameters: pH, Secchi disc transparency, temperature, alkalinity, total-P-concentration, conductivity, Fe-concentration, Ca-concentration, hardness (=CaMg-concentrations) and colour; totalP-concentration and transparency indicate different aspects of the lake´s nutrient status; colour and Fe-concentration provide information on the influence of humic substances; alkalinity and Ca-concentration provide data on, e.g., the result of liming measures; bioproduction increases with increasing water temperature (see e.g., Wetzel, 1975).

• The theoretical water turnover time (T; determined from data on the water transport of inflowing rivers, Q, and lake volume, V; T=V/Q); water discharge (Q) is estimated from the specific runoff (see, Report from the National Swedish Board of Fishery, No. 1, 1982).

• Morphometric data, e.g., volume, lake area, mean depth, are determined from bathymetric maps (see Håkanson, 1981).

• Drainage area parameters, e.g., soil types, bedrocks, land use, lake distribution (see Appendix 2).

In 41 lakes (Fig. 2.4) in the counties of Västernorrland and Gävleborg, that suffered a relatively large fallout, we have collected data on the lake load of Cs-137, environmental factors (morphometry, water chemistry, drainage area characteristics), Cs-concentration in fish and remedial measures.

Trophic level	Primary prod. (g C/m2*yr)	Secci d. transp. (m)	Chloro-phyll-a$ (mg/m3)	Algal volume$ (g/m3)	Tot.-P$$ (mg/m3)	Tot.-N$$ (mg/m3)	Dominant fish
Oligot.	<30	>5	<2	<0.8	<5	<300	Trout, WF
Mesot.	25-60	3-6	2-8	0.5-1.9	5-20	300-500	WF, Perch
Eut.	40-200	1-4	6-35	1.2-2.5	20-100	350-600	Perch,Roach
Hypert.	130-600	0-2	30-400	2.1-20	>100	>1000	Roach,Bream

$ = Mean value for the growing period (May - Oct.)
$$ = Mean value for the spring circulation
WF = White fish

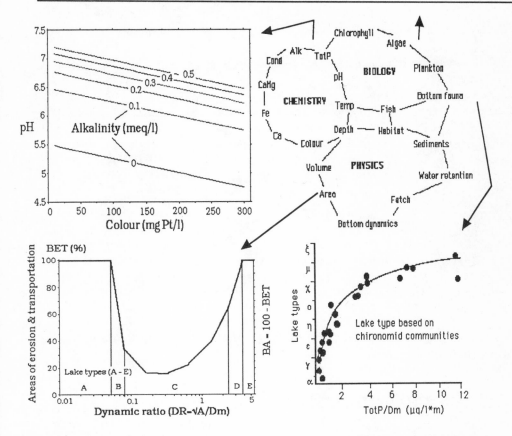

Fig. 2.3. Illustration of the complex interactions between various chemical, biological and physical factors that may be used to characterize a lake ecosystem and influence its sensitivity to toxic substances like mercury and caesium-137. Example:
Top. Characteristic features in lakes of different trophic status.
Centre left. The relationship between lake pH, alkalinity and colour from the equation:
$pH=6.94*(alk+0.01)^{0.0766} - 0.0025*Col + 0.63$. From Nilsson et al. (1989).
Lower right. The relationship between the ratio total-P/mean depth and the lake type as given by bottom faunistic criteria. From Saether (1979).
Lower left. The relationship between bottom dynamic conditions (the bottom area of erosion plus transportation, BET, or the area of fine sediment accumulation, BA), the lake type according to sedimentological criteria and the lake morphometric characteristics (the dynamic ratio, DR, between the square root of the lake area, √a, and the mean depth, Dm). Figure redrawn from Håkanson et al. (1990b).

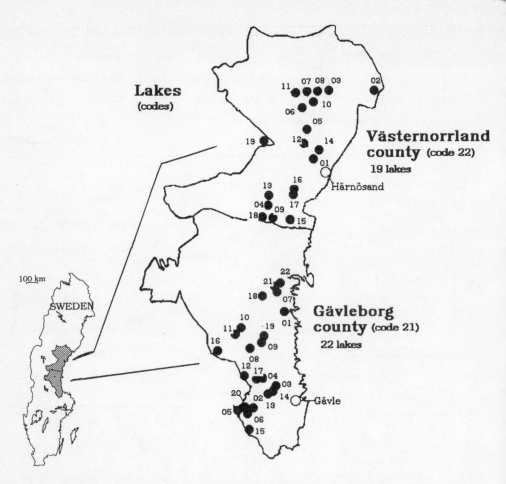

Fig. 2.4. The 41 lakes in the provinces of Gävleborg and Västernorrland and their codes.

2.2. LAKE CHARACTERISTICS

A presentation of relevant data from all our lakes is given in Appendix 1. For a more detailed description of the lakes and their drainage areas, see Bengtsson et al. (1987, 1988) and Johansson et al. (1987, 1988). Basic data on the lakes (lake number, name, x and y coordinates in the National Grid, i.e., N-S and E-W coordinates, lake area, pH, alkalinity, conductivity, colour, total-P, iron and hardness) are given in Table 2.1. It must be emphasized that Table 2.1 has been included here to provide background data on the conditions before the remedial measures were introduced (the table gives annual mean values for the period March 1986 - February 1987). Statistics (mean values, standard deviations, median values, min.

and max. values) for mainly the water chemical data in Table 2.1 have been compiled in Table 2.2.

Tables 2.1 and 2.2 illustrate, e.g., that our lakes generally are relatively acidic (mean pH=6), have low alkalinity (mean alk=0.07), have a relatively low bioproduction (totP=9.7, on average) and a relatively high colour value (about 100 units, on average). All these characteristics are typical for acidified, low-productive Swedish forest lakes. This is, thus, the typical project-lake but the variation between the lakes is, as can be seen from Table 2.2, significant.

2.3. METHODS

The accuracy in the determination of caesium in water depends on the entire handling process, from sampling to measurement (see Håkanson et al., 1988). The statistical uncertainty in the actual measuring process is determined by the detector efficiency, the background intensity and the time taken to conduct the measurement. The measurement method used gives a standard deviation of about 0.015 Bq/l, more or less regardless of sample activity.

The dry matter content for fish muscle was determined in order to allow calculation of Cs-concentrations in fresh material (dw%=100*dry weight/ fresh weight). The concentration of caesium was determined by gamma analysis behind a 10-cm lead shield and is given in Bq/kg dry weight or wet weight (the conversion factor, unless otherwise stated, is 0.25). In most cases, the measuring error has been kept below 5%. The measuring time varied from about 30 min to 24 h depending on the amount of Cs in the sample. The detection limit also depends on the measuring time and the size of the sample (the detection limit decreases with increasing fish weight). Thus, e.g., the detection limit is 500-1000 Bq/kg dw for individual fish samples and sediment trap material with a dry weight of about 0.25-0.3 g and a measuring time of around 15 h.

The following caesium parameters will be discussed:

• Cs-soil=fallout of Cs-137 in Bq/m^2 (determined for each drainage area from Fig. 2.1).

• Cs-wa=caesium in water samples (in Bq/l). If the mean value for a given year is used, this is indicated by, e.g., Cs-wa87 or Cs-wa88. If there are divergences from this nomenclature, special mention is made. The same coding is used also for the water chemical parameters, e.g., pH87 stands for the mean value of data from 1987. It should be emphasized that the

Table 2.1. Lake data and water chemical data before introduction of remedial measures (annual mean values 8603-8702).

A. Gävleborg county

Lake	Name	x	y	A	pH	alk	cond	Col	totP	Fe	CaMg
2101	Alebosjön	681231	156267	0.86	6.2	0.04	4.3	145	11.3	878	0.25
2102	Fäbodsjön	672210	153359	0.35	6.2	0.04	4.3	145	11.3	878	0.25
2103	Långsjön	673893	155114	0.74	6.2	0.05	3.2	95	10.9	1540	0.22
2104	Mångeln	674761	154550	1.75	6.1	0.04	2.6	131	8.8	1011	0.19
2105	Holmsjön	671962	151938	0.16	6.0	0.05	3.1	143	17.1	1200	0.24
2106	St Kröntjärn	671832	152967	0.11	6.2	0.06	3.7	140	10.3	728	0.27
2107	Redsjösjön	683036	155556	1.32	6.4	0.05	2.5	57	5.0	159	0.18
2108	Ramen	677527	153059	0.82	6.1	0.03	2.2	96	7.6	290	0.15
2109	Långsjön	678130	153944	0.22	5.8	0.02	2.5	109	8.4	633	0.15
2110	Ecklingen	679481	152245	0.44	6.6	0.17	3.8	112	11.8	472	0.31
2111	L:a Öjungen	679046	151727	0.46	6.4	0.08	2.7	83	6.4	290	0.18
2112	Stensjön	675186	152562	0.23	5.6	0.02	2.2	189	11.7	1705	0.18
2113	Vällingen	673387	155022	0.54	5.7	0.03	2.9	201	14.2	1340	0.23
2114	Järvsjön	673784	155146	0.35	6.2	0.08	4	153	13.3	1438	0.30
2115	Mörsen	670126	153202	0.13	6.2	0.09	3.6	153	13	895	0.32
2116	S:a Glittern	677206	150218	0.95	5.9	0.02	2	104	6.9	685	0.14
2117	Tansen	674863	153845	0.59	5.6	0.02	2.3	130	10.6	622	0.18
2118	N Maskinsj.	682510	154116	0.38	6.0	0.04	2.4	112	7.5	1995	0.21
2119	Bältbosjön	678588	154232	0.15	5.9	0.02	2	79	8.9	334	0.12
2120	Långsjön	671974	152544	0.16	6.2	0.06	3.6	133	8.5	577	0.28
2121	Blacksåstj.	683789	155174	0.12	6.4	0.11	3.1	115	11.6	1008	0.26
2122	Bottentjärn	683977	155509	0.16	5.8	0.03	2.2	110	10.0	651	0.16

B. Västernorrland county

lake	Name	x	y	A	pH	alk	cond	Col	totP	Fe	CaMg
2201	Selasjön	696898	159243	0.25	5.2	0.01	2.4	122	9.5	439	0.14
2202	Kittelsjön	702300	166018	0.40	6.0	0.04	3.0	42	5.0	150	0.17
2203	Galasjön	703317	161623	1.7	5.9	0.03	2.3	36	3.0	50	0.13
2204	Huljesjön	692272	155761	1.27	6.2	0.07	2.8	69	6.5	283	0.20
2205	Lövsjön	699031	158799	1.06	6.5	0.11	3.9	56	8.0	240	0.28
2206	Rävsjön	702629	158342	0.60	5.6	0.03	2.1	128	11.9	770	0.14
2207	Lövsjön	703846	158935	2.04	5.4	0.01	1.6	63	5.0	248	0.09
2208	Uvsjön	703399	169792	1.33	5.9	0.04	2.4	73	6.0	305	0.16
2209	Bysjön	690187	156365	0.63	6.2	0.07	2.6	95	10	405	0.19
2210	Oppsjön	702389	159908	1.58	5.2	0.0	1.8	120	12	463	0.10
2211	Hermansjön	703946	158439	2.8	6.1	0.05	2.0	58	6.0	228	0.13
2212	S.-Habborn	697633	158415	0.28	6.0	0.06	2.4	79	8.7	405	0.18
2213	V.-Lövsjön	693032	155932	0.47	6.4	0.08	3.0	38	7.1	166	0.22
2214	Lill-Selss.	697055	159312	0.07	5.1	0.01	2.3	115	10.1	499	0.14
2215	Herrbodtj.	690012	156950	0.24	6.2	0.07	2.8	70	9.5	269	0.19
2216	Lill-Bands.	693065	157509	0.16	6.5	0.22	5.5	35	11.	160	0.42
2217	Hamstasjön	692930	157589	0.18	6.6	0.37	8.2	46	22.9	357	0.64
2218	V.-Långeds.	690618	155477	0.25	6.5	0.24	4.6	69	13.8	459	0.37
2219	Ödingen	698758	155293	2.1	5.9	0.03	1.9	112	7.5	438	0.14

x: X-coordinate y: Y- coordinate A: Total lake area (km^2) alk: Alkalinity (meq/l) CaMg: Hardness (meq/l) cond: Conductivity (mS/m) Colour in (mg Pt/l) totP: Concentration total-phosphorus (µg/l) Fe: Iron concentration (µg/l)

Table 2.2. Statistical presentation of lake data for the 41 lakes.

Parameter	Mean MV	Std. dev. SD	Median	Min.	Max.
lake area (km^2)	0.69	0.66	0.44	0.07	2.8
pH	6.03	0.38	6.1	5.1	6.6
alkalinity (meq/l)	0.066	0.071	0.04	0	0.37
conductivity (mS/m)	3.0	1.12	2.6	1.6	8.2
colour (mg Pt/l)	101	41	109	35	201
totP (µg/l)	9.7	3.6	9.5	3	22.9
Fe (µg/l)	626	462	463	50	1995
CaMg (meq/l)	0.22	0.1	0.19	0.09	0.64

caesium determination of lake water samples involves a total analysis and that most of this caesium is probably bound to particles (see Broberg and Andersson, 1989).

• Cs-su and Cs-bo=caesium in material from sediment traps placed near the surface and near the bottom (Bq/kg dw). We discuss data (lake means) from 1986, 1987 and 1988.

• Cs-pi=caesium in pike (Bq/kg wet weight, ww). We use data from individual determinations in 1987 and from pooled samples for entire lakes in 1988.

• Cs-pe=caesium in 1+perch (i.e., 1-year-old perch fry; values in Bq/kg ww). We discuss data from 1986, 1987 and 1988 from pooled samples for entire lakes.

Table 2.3 summarizes results (means, standard deviations, min. and max. values) for caesium parameters from 1987. Unless otherwise stated, we have used data according to the following:

• Sediment traps: Traps deployed in June after the peak of the spring flood, recovery in September in connection with fishing for perch; on average, the traps have been in position for about 100 days (standard deviation 5 days).

• Pike: catching period from second half of January until late June.

• Perch: catching period from about July 15 until about October 15, generally during September.

• Water sampling from May to September.

The entire analytical programme has been carried out in 15 lakes. For the rest of the lakes, there are no caesium data from lake water and sediment trap samples, and no data on totP, Fe, Ca, K, Secchi depth and water temperature.

Table 2.3. Statistics for the caesium parameters from 1987 for our 41 lakes. Note that data from sediment traps and lake water are only available for 15 lakes.

Parameter	n	Mean MV	Std. dev. SD	Median	Min.	Max.
Cs-pi87 (Bq/kg ww)	41	2392	353	1355	195	7310
Cs-pe87 (Bq/kg ww)	41	5036	6015	2650	129	24475
Cs-soil (kBq/m^2)	15	25.2	20.8	15.0	2.5	70.0
Cs-su87 (Bq/kg dw)	15	6150	4989	5600	500	16953
Cs-bo87 (Bq/kg dw)	15	8455	7838	5312	410	23145
Cs-wa87 (Bq/l)	15	0.314	0.265	0.160	0.030	0.790

As can be seen from Table 2.3, the Cs-concentrations in fish, water and the sediment trap material do not follow a normal distribution (this also applies to fallout). In a special section, we will discuss questions concerning the statistical distributions of the parameters and the importance this may have on ecometric models.

The data have been handled using the standard programmes Excel (Macintosh) and Lotus (IBM), statistical calculations have been made using Statgraphics (IBM), Statview and JMP (Macintosh), dynamic modelling using Labview (Macintosh) and graphics with MacDraw, SuperPaint and Delta-Graph (Macintosh).

3. LOAD DIAGRAMME

The objective of this chapter is to follow and discuss the various steps in the ecometric analysis to establish load models and load diagrammes.

3.1. ECOMETRY — PRESUPPOSITIONS

Håkanson (1984a, 1990) gives a more detailed presentation of the theoretical basis required in obtaining load diagrammes (effect, dose, sensitivity) using ecometric analysis for environmetally hazardous substances in general. Many factors may have an influence on how an effect parameter varies between lakes. The ecometric analysis aims at identifying the most important factors in this respect. Frequently, there are no **causal** explanations to phenomena that can be established **statistically**. One of the advatages of the ecometric approach is then that it provides a possibility to rank factors exerting influence on an effect parameter so that future research resources can be concentrated on these factors. Naturally, **when using ecometric analysis on the ecosystem level** (i.e., for entire lakes, coastal areas, etc.), **it is not possible to explain many phenomena that occur on the individual, organ or cell levels**.

3.1.1. REPRESENTATIVE DATA

The intention in this section is simply to emphasize that sampling and analytical techniques always embody a certain degree of error, which is different for different parameters (see Håkanson et al., 1990c).

Figure 3.1A shows the relative standard deviations (i.e., $V=100*SD/MV$; SD=standard deviation; MV=mean value) for caesium (in pike, Cs-pi and in lake water, Cs-wa) and the water chemical parameters Fe, Ca, totP, pH, conductivity, hardness (Ca+Mg) and colour. The figure illustrates that there are large general, relative differences in analytical reliability for these parameters. The error (=relative standard deviation) for the caesium measurement in lake water is, however, not a constant but varies depending on the Cs-concentration of the lake water samples. The value given in the figure for Cs-wa is based on the standard deviation being largely equal to 0.015 Bq/l regardless of sample activity (see Håkanson et al., 1988). The value 5% is related to the mean value for Cs-wa and Cs-pi for 1987 (data from Table 2.3).

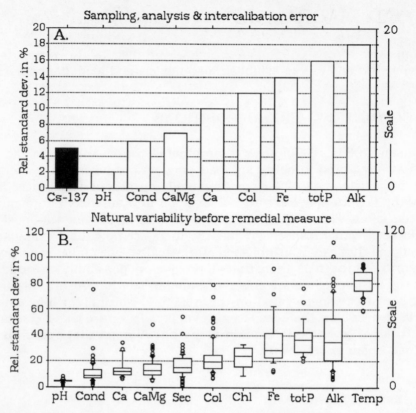

Fig. 3.1A. Compilation of relative standard deviations (V in %) for Cs-317 (in fish, sediments and water), pH, conductivity, hardness (CaMg), Ca, colour (Col), Fe, total-P and alkalinity linked to limitations/errors in analytical techniques, sampling and laboratory determinations (including intercalibrations).
B. Relative standard deviations for different water parameters (including water temperature). The values are based on data from 75 Swedish lakes. From Håkanson et al. (1990b).

From Fig. 3.1A, it can also be seen that pH (which by definition is a logarithmic value) can be determined with a high reliability (2%) and that Fe and totP generally give a much lower reliability (about 15%). The reliability, or the error, with which one can determine a parameter, influences the **degree of explanation** (=coefficient of determination=r^2-value) that can be obtained in the ecometric model. Thus, it is important to know that when parameters such as colour, Fe-content and totP are used, the chances to obtain models that yield high r^2-values compared to empirical data are relatively poor.

However, the analytical error is only one consideration in this respect. This is shown in Fig. 3.1B. Here, we can see that for data from **an entire**

year within natural **lakes** there are sometimes extremely high V-values. It is not surprising that the water temperature varies during a year — the V-value is 80%, but it is remarkable that standard parameters such as colour, totP, Fe, and alkalinity give V-values ranging between 20 and 40% on average in this context. This implies that one must analyse many samples in order to obtain lake-typical values with a given statistical reliability for these parameters. The reason for the high V-values is partly the analytical problems, but the natural variations in these parameters in the lakes due to the fact that they take part in chemical, physical and biological reactions is an equally important reason.

3.1.1.1. The sample formula

Table 3.1 gives fish statistics from the four lakes where we have analysed most fish. The table shows that the relative standard deviation (V) for Cs-pi is 32.8%, on average, which is a relatively high value. Thus, many pike must be analysed to establish a lake-typical mean value for Cs-pi.

Table 3.1. Fish statistics (mean values of Cs-pi87 in Bq/kg ww, MV; standard deviations, SD; relative standard deviations, V; and the number of fish analysed, n).

Lake	MV (Bq/kg ww)	SD (Bq/kg ww)	V %	n
2107, Redsjösjön	3483	1106	31.8	14
2110, Ecklingen	322	74	23.0	15
2201, Selasjön	4654	1821	39.1	15
2207, Lövsjön	6813	2534	37.2	15
			Mean: 32.8	

A general formula stating how many samples are required (n) in order to establish a lake mean value is (see Håkanson, 1984b):

$$n = t*V/y^2 + 1 \qquad (3.1)$$

where y=the error accepted in the mean value; y=1 implies 100% error; subsequently, we will determine the mean value with 95% certainty (p=0.05), which gives a t-value of 1.96. This implies that Eq. (3.1) can be written:

$$n = 1.96*V/y^2 + 1 \qquad (3,2)$$

The relationship between n, V and y is illustrated graphically in Fig. 3.2. If the relative standard deviation is 32.5, which is almost equal to the mean value for Cs-pi for the four lakes in question, we can see from Fig. 3.2 that about 40 pike are required to establish a lake-typical mean value provided that we accept an error of 10% (y=0.1). Since in most lakes we have only analysed about 5 fish, one can calculate for n=5 and V=32.5 from Eq. (3.2) that this gives an error of about 33%. Consequently, we must expect that our empirical values only give a fairly rough measure of the lake-typical mean value for Cs-pi. And the same type of argument could be made for **any given parameter.**

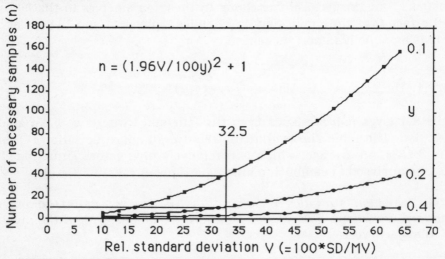

Fig. 3.2. Nomogram showing how many samples (e.g., pike) must be analysed (n) in order to establish a lake-typical mean value with a given statistical error (y) and a given statistical certainty (95%). V is the relative standard deviation in percent.

It is important to remember this uncertainty in the empirical data in the subsequent discussions when we will try to quantify the variation in Cs-pi **between** lakes and how this variation may be linked to different dose and sensitivity parameters. Since we do not have very reliable empirical data on Cs-pi, one cannot expect to obtain explanatory models which give a full 100% degree of explanation. The **uncertainty** in the empirical determinations of caesium in fish from lakes for which relatively few fish have been analysed may thus result in there being marked divergences when the model data are compared to empirical data. In such cases, occasional widely diverging values need not depend on errors and deficiencies in the models but may also depend on deficiencies in the empirical basic material.

3.1.1.2. Standardizations - caesium in pike of different weight

Can the variability of a given parameter **within** a lake be reduced by some sort of standardization? Is, e.g., the variability of Cs-pi dependent on the weight of the fish?

A total of 250 individual pike from 1987 have been analysed for Cs-137. From 1988, only pooled samples have been analysed. The analyses of pooled samples imply considerable cost advantages but naturally do not provide any base for statistical assessments of how lake-representative data can be obtained. Thus, the discussions in this section will only concern material from 1987. It is important to keep this in mind since it is probable that the relationship between the Cs-pi-values and the pike weight will change with time after Chernobyl. A basic question has been: Does the Cs-concentration vary with the weight of the pike? If the answer is yes — how should one account for this?

From lakes Tansen (2107), Ecklingen (2110), Selasjön (2201) and Lövsjön (2207) there are more than ten analysed pike per lake. Figure 3.3 illustrates how the Cs-concentration varies with weight in these four lakes. The figure illustrates that there is no clear relationship between the Cs-concentration and the weight of pike; in two of the lakes the regression line points weakly upwards, but the correlation coefficients (r-values) are low; in the two other lakes (Tansen and Selasjön) there is a significant (although weak) negative relationship between Cs-pi87 and weight. Table 3.2 gives the entire data material. The table gives the linear correlation coefficients (r) between Cs-pi87 and pike weight, the statistical significance (p), the number of analysed pike per lake (n), the mean weight of the pike and their mean concentration of caesium (Cs-pi87).

On the basis of the 1987 material, we must conclude that one cannot improve the representativity of the Cs-pi87-values much by standardizations for pike weight. If this is to be done, then it should be done for each lake individually, and on the basis of a larger data-set for each lake than we have at our disposal.

Figure 3.4 gives the relationship between the lake mean values for pike weight and the Cs-pi87-values (data from Table 3.2). The correlation coefficient is zero. Pike mean weight varies from about 500 to about 2000 g. This variation in weight does not appear to have any major influence on the Cs-concentration in pike this year.

Figure 3.5 illustrates how Cs-concentrations in individual pike varied during the first 6 months of 1987. The first fish were caught during the second half of January and the last before late June. Most fish were caught

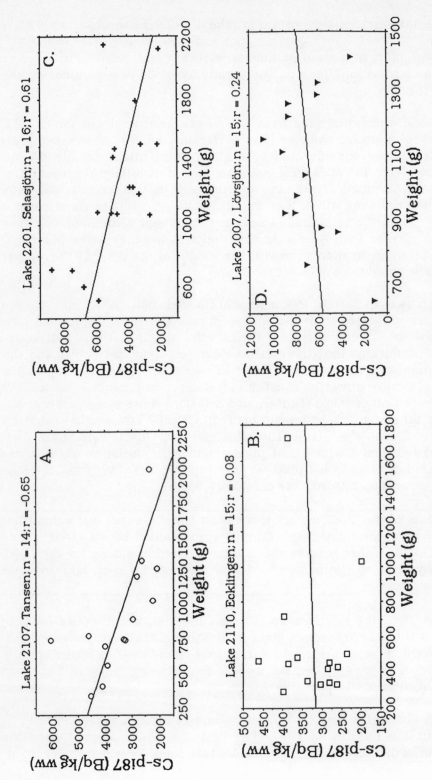

Fig. 3.3. Relationship between caesium concentration in pike in 1987 (Cs-pi87 in Bq/kg wet weight) and pike weight (g) in four lakes, (A) Tansen, (B) Ecklingen, (C) Selasjön and (D) Lövsjön. The figure shows the regression line, the number of individual caesium analyses (n) and the linear correlation coefficients (r).

Table 3.2. Correlations between pike weight and caesium concentration in pike in our 41 lakes; pike mean weights and lake mean values for Cs-pi87 in the lakes.

Lake nr	Corr. coeff. r	Certainty level p	Number pikes n	Mean weight (g)	Mean Cs-pi87 (Bq/kg ww)
2101	-0.82	0.09	5	1052	1158
2102	-0.59	0.3	5	1445	195
2103	0.46	0.43	5	628	1359
2104	0	1	5	930	1132
2105	-0.2	0.75	5	1235	531
2106	-0.47	0.43	5	1234	297
2107	-0.65	0.01	14	923	3483
2108	-0.77	0.13	5	1329	743
2109	-0.41	0.49	5	1629	417
2110	0.08	0.77	15	556	322
2111	-0.47	0.42	5	1372	258
2112	-0.9	0.04	5	1277	623
2113	-0.43	0.47	5	933	1784
2114	-0.99	0.01	5	1035	400
2115	-0.03	0.96	5	1964	285
2116	0.23	0.71	5	1139	537
2117	-0.32	0.61	5	941	942
2118	0.03	0.99	5	1838	1346
2119	-0.63	0.38	4	1136	821
2120	-0.77	0.01	10	1205	260
2121	-0.76	0.13	5	1135	2009
2122	-0.18	0.78	5	1328	1191
2201	-0.61	0.01	15	1253	1821
2202	-0.79	0.11	5	1668	5797
2203	-0.78	0.12	5	894	1215
2204	-0.8	0.1	5	1727	3016
2205	0.24	0.69	5	1180	1355
2206	-0.49	0.4	5	1361	1932
2207	0.24	0.39	15	1055	6813
2208	0.12	0.84	5	1047	5932
2209	0.12	0.85	5	1265	2194
2210	-0.77	0.13	5	1225	7310
2211	-0.46	0.43	5	1548	5919
2212	-0.82	0.09	5	1493	4147
2213	-0.82	0.09	5	1301	4444
2214	0.97	0.01	5	833	6940
2215	0.63	0.26	5	750	1683
2216	-0.63	0.26	5	992	2692
2217	0.54	0.35	5	940	769
2218	-0.12	0.85	5	751	818
2219	0.53	0.35	5	839	6483
MV	**-0.3**	**0.4**	•	**1180**	**2229**
SD	**0.5**	**0.32**	•	**318**	**2201**

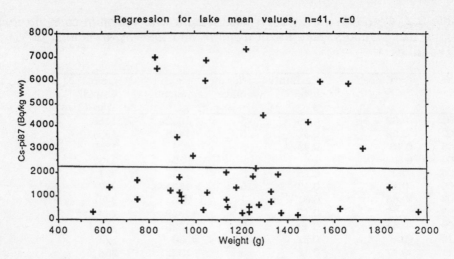

Fig. 3.4. The relationship between lake mean values for Cs-pi87 and pike weight. The figure shows the regression line, the number of lakes (n) and the linear correlation coefficients (r).

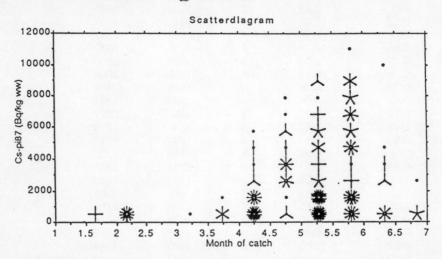

Fig. 3.5. Relationship between date of catch and Cs-pi87 for all individual pike (n=250). Each line in each grouping represents one analysis.

during April and May. Figure 3.5 might lead us to believe that there is a clear relationship between time of catch and Cs-concentration in fish, but this would be an incorrect conclusion. In order to reach such a conclusion, it is necessary to carefully study the conditions in each lake. Such a conclusion can **not** be reached from Fig. 3.5, which shows results from 41 lakes where the Cs-pi87 values vary widely between the lakes. However, Andersson et al. (1990) have shown that Cs-concentrations in pike from 1987 increased by up to ca. 50-75% after the production period (May-October) in relation to the time before the production period (Jan.-April).

This means that Fig. 3.5 only **indicates** a seasonal variation in Cs-pi87 and that the figure, in fact, mainly illustrates the time when the fish were caught.

The same results, as shown between Cs-pi87 and weight, would also be obtained if we related Cs-pi87 to fish length, since the relationship between pike length and weight is very strong (r^2=0.92). This is shown for all individual pike (n=25) in Fig. 3.6. Thus, weight and length (and age) form a **cluster**; these parameters are related to each other.

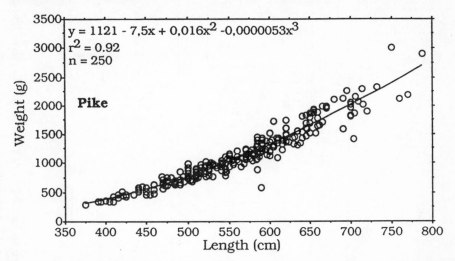

Fig. 3.6. Relationship between pike weight (g) and pike length (cm). The figure gives the regression line, its equation, the r^2-value and the number of pike analysed (n).

3.1.1.3. Impact of range and number of fish

Since the r^2-value is used as an important tool in this context, it is also relevant to discuss the statistical factors that influence the r^2-value, apart from the ecological/methodological factors. Admittedly, this work does not discuss the statistical aspects of ecometry, but **questions concerning the number of samples and their range concern both sampling strategy and data evaluation and have been considered of relevance here.**

Figure 3.7 shows five regressions between Cs-pi88 and Cs-pi87. The first, Fig. 3.7A, shows the relationship for the entire material (41 lakes). The problem illustrated by the following four figures concerns: How is the degree of explanation altered when the range decreases? In order to obtain comparability, the sample size has been kept constant at n=15. The 15 lakes included in the test have been randomly selected after establishing the range. Figures 3.7B-E show how the r^2-values successively de-

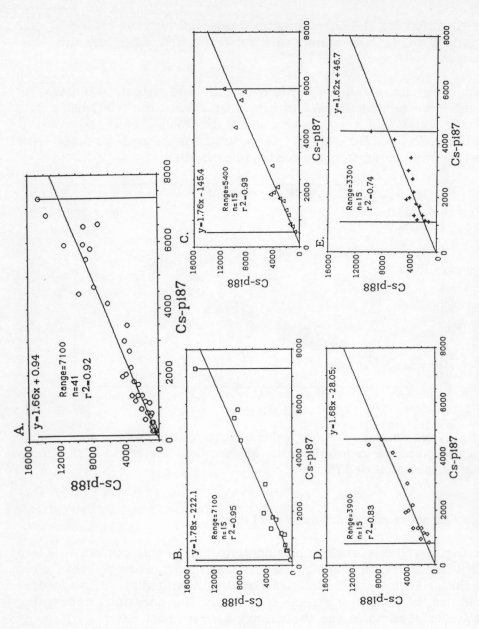

Fig. 3.7. Illustration of how the range influences the regressions between Cs-pi87 and Cs-pi88 (slope coefficient, intercept and r². value). A gives the entire data material (n=41). B-E have different ranges (n=15 in all examples).

crease from 0.95 to 0.76 when the range decreases from 7100 to 3300. The fact that the r^2-value in Fig. 3.7B is higher than for the parent population (i.e., the data in Fig. 3.7A) depends on coincidence.

> The range is thus an important factor for the r^2-value; the wider the range the higher the r^2-value (all other factors kept constant).

Figure 3.8 shows six regressions between Cs-pi87 and Cs-pi88 where the number of analyses are randomly and successively decreased from 30, 15, 5, 3 (twice) to 2. The figure shows that not only the r^2-value but also the slope coefficients (and intercepts) vary considerably when n decreases; the r^2-values vary between 0.72 and 1.00 and the slope coefficients between 0.75 and 1.9.

> The simple message is: The larger the n value, the more reliable the regression. The error, e.g., in the slope coefficient may randomly be very large at small n values; occasionally it is also possible to obtain very high r^2-values at low n. If we only have two analyses (n=2) the regression cannot be done since the r^2-value then will be 1.00. These are general rules that apply to all regression analyses.

Another example of the role of the range and the number of samples in regression analysis is given in Fig. 3.9A. This figure illustrates the relationship between mercury (Hg) in pike (Hg-pi) and "scattered" data on the pH of the lakes for the **entire** natural (Swedish) range, i.e., pH varies from about 4 to about 8 and Hg-pi from about 0.1 to about 3 (mg Hg/kg wet weight). We can see that there is a large spread around the regression lines for these data (894 lakes from all of Sweden; Fig. 3.9A). This is also the case for the data for 75 selected lakes giving a smaller range in Fig. 3.9B. There is, however, a clear general correlation: The Hg-pi values increase when pH decreases. There are several reasons for this relationship (see Håkanson et al., 1990b). Here, we will not discuss this aspect, but simply emphasize that if one uses material with a large inherent spread, any relationship may be obtained from a few samples with a limited range. This is illustrated by the bold regression line for the data marked with black dots in Fig. 3.9B. According to data from these 6 selected lakes, the Hg-pi values would **increase** markedly when pH increases! This is completely in contrast to reality, but a result that could be obtained from reliable empirical data.

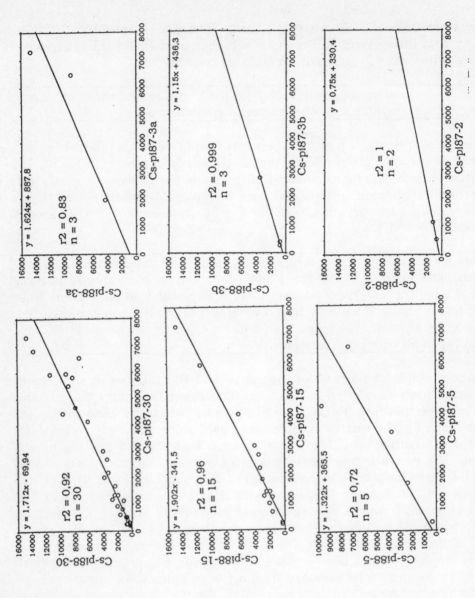

Fig. 3.8. Illustration of how the number of analyses influences the regression between Cs-pi87 and Cs-pi88 (slope coefficient, intercept and r²-value). The number (n) varies from 30, 15, 5, 3 (two diagrammes) down to 2.

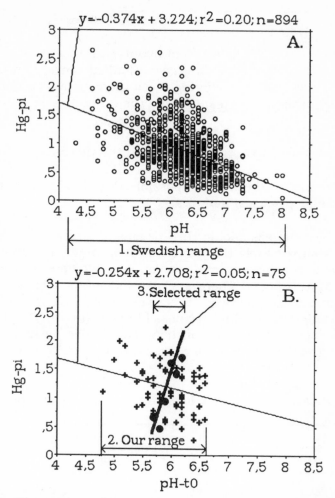

$$y = -0.374x + 3.224; r^2 = 0.20; n = 894$$

A.

$$y = -0.254x + 2.708; r^2 = 0.05; n = 75$$

B.

Fig. 3.9. Illustration of how relationships can be obtained between two parameters, mercury in 1 kg pike, Hg-pi in mg Hg/kg wet weight, and lake water pH, for three different data-sets, when primarily the range of the parameters and the number of analyses vary.

A. Relationship between Hg-pi and lake pH as determined from <u>individual or few</u> water samples. The data material covers 894 lakes from all of Sweden and the entire pH-register, i.e., from about 4 to about 8. The regression line and the r^2-value are given.

B. The same relationship for 75 lakes, which cover a narrower pH-range. The pH-values here are mean values from monthly samples for about 1 year. The r^2-value, the slope coefficient and the intercept are not the same in this material as for the entire Swedish material.

The figure also illustrates that totally misleading relationships can be obtained between these two variables if very few lakes are chosen, e.g., from a certain commune. The large black circles, and the bold regression line, are based on data from six lakes. From Håkanson et al. (1990b).

Consequently, it is important to remember that all environmental parameters vary within a certain range, e.g., the Hg-pi values in Sweden vary between 0.15 and 3, the pH-values from about 4 to 8, etc. If one selects a few lakes, e.g., from a commune, that give a limited range, and conduct a regression analysis without considering these fundamental conditions, one can obtain totally meaningless results from the general perspective.

3.1.2. HIGHEST DEGREE OF EXPLANATION — PARALLEL SAMPLES

One way of determining the highest possible degree of explanation is to compare two similar empirical samples.

The fish in our two samples should then be caught at the same time of the year and be of approximately the same weight in order to standardize for possible variations in catching time and fish size. We have at our disposal sufficient pike from 5 lakes. There are also individual analyses of perch from 5 lakes and using these two sources (see Table 3.3), we can obtain a fairly good indication of the highest possible r^2-value for Cs-pi as effect parameter. The degree of explanation between the five mean values (MV) for both pike and perch was 0.97; for the entire material (n=10) the result was $r^2=0.96$; see Fig. 3.10.

Table 3.3. Comparison between two lake samples of pike and perch fry.

			Pike					
		Sample 1			**Sample 2**			
Lake	n	MV	SD	V	n	MW	SD	V
2107	6	3356	1477	44.1	8	3578	831	23.2
2110	7	337	77.5	23.0	8	309	73.7	23.8
2120	5	318	81.9	25.8	5	202	66.9	33.1
2201	9	3827	1592	41.6	8	5072	1295	25.5
2213	8	6795	2176	32.0	7	6833	3075	45.0
			Perch					
2105	8	847	496	58.6	8	776	727	93.7
2108	7	1596	211	13.2	6	2347	973	41.5
2118	4	8594	2282	26.6	4	11238	5348	47.6
2204	8	3758	1102	29.3	8	3094	1247	40.3
2213	8	7564	2598	34.4	8	10275	4773	46.5

The conclusion from these results is that, under the conditions prevailing with regard to the number of fish, the number of lakes investigated, the reliability of the caesium analysis, etc., we can maximally attain a degree of explanation of about 96% for Cs-pi as effect parameter.

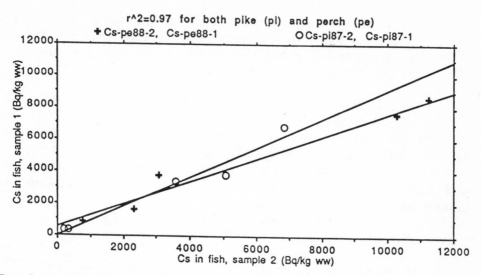

Fig. 3.10. The relationship between caesium concentration in fish from two lake samples (sample 1 on the y-axis and sample 2 on the x-axis); based on individual caesium analyses of pike and perch from 1987 and 1988.

This means that the **unexplained residual term** (RES) may be written as:

$$RES = 100*(1-r^2)$$

(3.3)

For $r^2 = 0.96$, RES is equal to about 4% [100*(1-0.96)].

3.1.3. "OUTLIERS"

Values diverging from the "normal" are called here "outliers". In this section, extreme values are discussed with examples concerning caesium in 1+perch.

Figure 3.11A shows the relationship between lake mean values for Cs-pe86 and Cs-pe87 for all 40 lakes for which we have data for these 2 years. The figure gives the regression line and the r^2-value, which is 0.37. This indicates a fairly weak relationship between the two parameters. However, it is obvious that there is a clear relationship between Cs-pe86 and Cs-pe87 and that the relatively low r^2-value depends on an extreme outlier — the value of 524 000 Bq/kg ww from lake 2211 in 1986. If this value is eliminated and the regression is recalculated, we get the result shown in Fig. 3.11B. The degree of explanation then increases to 63%; the slope and intercept of the regression line are quite different. We can now also see that two more values diverge markedly from the regression line.

This concerns data from lake 2213 (for 1986) and from lake 2207 (1987). Figure 3.11B reflects the normal relationships between Cs-pe86 and Cs-pe87 in a better way than Fig. 3.11A. If we also remove the two outliers in Fig. 3.11B, we get the result shown in Fig. 3.10C.

The r^2-value now increases to 0.89. The slope coefficient and the intercept are markedly changed and a further outlier is revealed, namely lake 2202. When this value is removed, we get an r^2-value as high as 0.95 (Fig. 3.11D). We can also see from Fig. 3.11D that there is no point in removing further data, that **the most relevant slope coefficient between the two years is 1.5, that the intercept is about 2000, and that the regression line is clearly above the line y=x**. The values for Cs-pe for 1986 are thus throughout higher than the values for 1987, i.e., the values decrease with time after Chernobyl.

Figure 3.11E shows the relationship between Cs-pe87 (n=39; lakes 2202 and 2207 eliminated) and Cs-pe88. Here, there are no clear outliers. The degree of explanation is 85%. The slope coefficient is about 2, which suggests that the average decrease from 1987 to 1988 is of the same magnitude but is slightly faster than the decrease between 1986 and 1987.

If we want to find how the concentrations co-vary generally between lakes, it is important to eliminate outliers. It should, however, be emphasized:

☞ **that outliers may only be eliminated if they can be demonstrated to be erroneous,**

☞ **that one should try to find out why the values are abnormal,**

☞ **that true outliers may provide valuable knowledge.**

As regards the outliers in Fig. 3.11, lakes 2202, 2207, 2213 and 2211, we can note:

• that the variation in Cs-pe was very large in both 1986 and 1987 (see Håkanson et al., 1988). It would, e.g., require analyses of, on average, 62 individual perch from each lake to establish a lake-typical mean if we accept an error in the mean value of 10% (with 95% certainty). The corresponding figure for pike is about 40 fish. However, we have only analysed pooled samples of perch from these "outlying lakes" and thus cannot provide data from individual fish analyses. We have only analysed 10-15 perch per lake which thus provide space for considerable errors in the determination of the lake means for Cs-pe.

• that the extreme value from lake 2211 in 1986 probably depends on the analysis including a "hot particle".

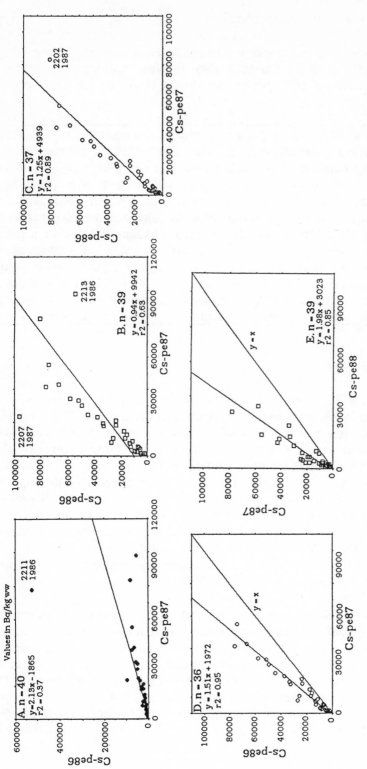

Fig. 3.11. A. Relationship between Cs-pe86 and Cs-pe87. Marking of the first "outlier", lake 2211. Analytical values from 40 lakes are compared. The degree of explanation is 39%.
B. Relationship between Cs-pe86 and Cs-pe87 when the "outlier", lake 2211, has been eliminated. Analytical values from 39 lakes are compared. The degree of explanation is 63%. Two new "outliers", lakes 2207 and 2213, have been marked.
C. Relationship between Cs-pe86 and Cs-pe87 when three "outliers" have been eliminated. Analytical values from 38 lakes are compared. The degree of explanation is 89%. A new "outlier", lake 2202, has been marked.
D. Relationship between Cs-pe86 and Cs-pe87 when four "outliers" have been eliminated. Analytical values from 39 lakes are compared. The degree of explanation is 95%. The y=x line has been marked.
E. Relationship between Cs-pe87 and Cs-pe88. Analytical values from 39 lakes have been compared. No clear "outliers" can be identified. The degree of explanation is 85%. The y=x line has been marked.

3.1.4. TIME- AND AREA-COMPATIBLE DATA

Many dose and sensitivity parameters can influence the value of the effect parameter. Many of these parameters vary in time and space, whereas others are constant (in the time scales of interest if we use Cs-pi as an effect parameter).

> It is important to establish the representativity of the samples in time and space. Here, the fundamental point is what the effect parameter represents. The pike may come from a limited area but the pike´s prey and the food eaten by the prey reflect the Cs-contamination over a much larger area than the area around the place of catch. The caesium entering the pike is metabolized and the Cs-concentration in pike has a biological half-life of about 1 year (Carlsson, 1978). But caesium in lake pike (after a point discharge such as that in the Chernobyl event) also has an ecological or true half-life (see Chap. 5.4) that depends on many variable factors, e.g., the Cs-concentrations in food organisms, lake water and sediments. Subsequently, a few examples will we given to point out that it is important to establish the time and area constants of the given parameters.

3.1.4.1. Caesium in pike

Caesium in pike 1988 (Cs-pi88) refers to pike caught during the spring spawning period in 1988, mainly during April, May and June. The caesium present in the lake after the fish has been caught cannot, of course, enter the fish. But for what period **prior** to the catch should one determine the mean value (or similar value) for dose and sensitivity parameters to obtain a maximum of compatibility? Is the dose and water chemistry during the months prior to catching most relevant or does it concern a 6-month period, a year or several years prior to catching? If we base our calculations on the r^2- or r-value, what period should one then use to obtain a mean value so that the r^2-value is as high as possible? And in such a case, why have we reached such a decision? These are major questions in establishing time-compatible data.

Table 3.4 shows a correlation matrix intended to illustrate the problem. In this case, the mean values have been determined for the following water chemical parameters: pH, conductivity, totP and hardness (CaMg), which we know are of importance for caesium uptake in fish (see Håkanson et al., 1988). The mean values of these parameters have been determined for the following time periods: The entire period from 1986 until September 1988 (the values are written as, e.g., pH-86-88); the period from 1987 until September 1988; the period from 1/7-87 until September 1988, the entire 1988 up to and including September; and the 6-month period after the pike have been caught. These periods form a ladder. Table 3.4 shows:

Table 3.4. A correlation matrix to illustrate the co-variation (time compatibility) between the effect parameter Cs-pi88 and four water chemical parameters (pH, conductivity, totP and hardness), when these are determined for five different time periods.

	Cs-pi88	pH-86-88	pH-87-88	pH-7-87/88	pH-88	pH-5-9-88
Cs-pi88	1,00	-0,54	-0,55	-0,26	-0,07	0,09
pH-86-88	-0,54	1,00	0,95	0,86	0,65	0,55
pH-87-88	-0,55	0,95	1,00	0,93	0,74	0,61
pH-7-87/88	-0,26	0,86	0,93	1,00	0,91	0,80
pH-88	-0,07	0,65	0,74	0,91	1,00	0,93
pH-5-9-88	0,09	0,55	0,61	0,80	0,93	1,00
cond-86-88	-0,50	0,74	0,74	0,64	0,41	0,33
cond-87-88	-0,49	0,76	0,80	0,73	0,52	0,43
cond-7-87/88	-0,39	0,75	0,81	0,81	0,65	0,56
cond-88	-0,35	0,72	0,80	0,82	0,74	0,66
cond-5-9-88	-0,19	0,64	0,72	0,80	0,77	0,74
totP-86-88	-0,23	0,11	0,09	-0,01	-0,17	-0,22
totP-87-88	-0,10	0,09	0,07	0,01	-0,11	-0,16
totP-7-87/88	-0,08	0,05	0,04	-0,02	-0,15	-0,19
totP-88	-0,05	0,02	0,00	-0,03	-0,15	-0,18
totP-5-9-88	-0,12	0,00	0,01	-0,03	-0,17	-0,19
CaMg-86-88	-0,58	0,75	0,77	0,64	0,41	0,32
CaMg-87-88	-0,55	0,75	0,82	0,73	0,54	0,44
CaMg-7-87/88	-0,41	0,73	0,82	0,81	0,68	0,58
CaMg-88	-0,36	0,68	0,78	0,80	0,76	0,67
CaMg-5-9-88	-0,16	0,57	0,67	0,76	0,78	0,75

• That the parameters from the longest period generally show the highest r-values with regard to Cs-pi88. This may be surprising, but probably mainly depends on the fact that comprehensive measures (liming, fertilization, etc) were introduced during 1987. These data show that the measures had an quick and significant influence on water chemistry more than on the Cs-concentration in pike. There are, as expected, no relationships between Cs-pi88 and any of these parameters for the period immediately following the catching, e.g., the r-value between Cs-pi88 and pH-88 is -0.07.

• That totP generally shows low r-values with regard to Cs-pi88.

• That pH, conductivity and CaMg, as expected, follow each other.

If the situation had been such that the ecological half-life for Cs-137 in lake pike had been short, let us say a few months or less, then the highest r-values should have been between Cs-pi88 and the values dominated by the data from 1987. This does not appear to be the case.

3.1.4.2. Caesium in 1+perch

The question is: Are there better correlations between the sensitivity parameters from the period before the catching and this effect parameter? Perch from 1988 were caught in August and September. Consequently, we have used the same division into time periods as for the pike and have also added two periods: The bioproduction period from 1988 and data from August 1988 (CaMg-87-88). The results are summarized in Table 3.5.

It can be seen from the table that also for this effect parameter the best correlation to the data is for the longest periods and the lowest r-values appear for the short periods immediately before catching. The result indicates that the remedial measures from 1987 may have been rather ineffective to reduce caesium in fish. It may also imply that the measures will have a positive effect but that it is still too early for this to be demonstrated, i.e., that the ecological half-life for Cs in fish is relatively long and/or that the equilibria that influence the distribution of caesium in different chemical forms and thus the biological uptake of caesium take a relatively long time to achieve.

3.1.4.3. Dose and sensitivity parameters

The water transport to the lakes is normally greatest during spring and autumn. Figure 3.12 has been included to remind us that all water chemical parameters and all dose parameters vary in time in one and the same lake, and also to emphasize that we have introduced comprehensive remedial measures that affect water quality without having anything to do with the normal inflow/precipitation. This figure shows how pH varies in a randomly selected lake (lake 2201). There are considerable pH variations ranging from 5.7 to 7.4. **It is not obvious which value from such a frequency distribution should be chosen as the most representative for the lake and most compatible with the Cs-concentration in pike**.

All parameters also vary between the lakes. In this work, the focus is on the values that represent the lakes since our interest mainly concerns differences between lakes. Figure 3.13 gives an example of how the mean pH for 1986, i.e., before the introduced remedial measures, varied between our lakes.

Table 3.5. A correlation matrix to illustrate the co-variation (time compatibility) between the effect parameter Cs-pe88 and four water chemical parameters (pH, conductivity, totP and hardness), when these are determined from seven different time periods.

	Cs-pe88	pH-86-88	pH-87-88	pH-7-87/88	pH-88	pH-5-9-88	pH-6-8-88	pH-8-88
Cs-pe88	1,00	-0,59	-0,61	-0,35	-0,18	0,01	0,08	0,12
pH-86-88	-0,59	1,00	0,96	0,87	0,68	0,57	0,44	0,25
pH-87-88	-0,61	0,96	1,00	0,93	0,76	0,62	0,52	0,33
pH-7-87/88	-0,35	0,87	0,93	1,00	0,92	0,80	0,72	0,53
pH-88	-0,18	0,68	0,76	0,92	1,00	0,93	0,86	0,69
pH-5-9-88	0,01	0,57	0,62	0,80	0,93	1,00	0,94	0,77
pH-6-8-88	0,08	0,44	0,52	0,72	0,86	0,94	1,00	0,87
pH-8-88	0,12	0,25	0,33	0,53	0,69	0,77	0,87	1,00
cond-86-88	-0,50	0,71	0,68	0,59	0,41	0,34	0,25	0,11
cond-87-88	-0,52	0,74	0,75	0,68	0,51	0,42	0,34	0,20
cond-7-87/88	-0,47	0,75	0,78	0,76	0,63	0,54	0,47	0,32
cond-88	-0,44	0,73	0,77	0,77	0,70	0,62	0,56	0,41
cond-5-9-88	-0,31	0,68	0,72	0,78	0,75	0,71	0,66	0,51
cond-6-8-88	-0,31	0,65	0,70	0,76	0,74	0,70	0,66	0,53
cond-8-88	-0,33	0,63	0,70	0,73	0,68	0,65	0,61	0,48
totP-86-88	-0,20	0,22	0,17	0,07	-0,07	-0,12	-0,19	-0,12
totP-87-88	-0,08	0,17	0,13	0,07	-0,06	-0,11	-0,14	-0,06
totP-7-87/88	-0,08	0,16	0,11	0,05	-0,08	-0,12	-0,14	-0,07
totP-88	-0,04	0,14	0,09	0,05	-0,07	-0,10	-0,12	-0,05
totP-5-9-88	-0,07	0,12	0,09	0,05	-0,09	-0,12	-0,13	0,01
totP-6-8-88	-0,04	0,09	0,10	0,08	-0,03	-0,06	-0,06	0,09
totP-8-88	-0,08	0,14	0,11	0,07	-0,05	-0,13	-0,12	0,06
CaMg-86-88	-0,57	0,75	0,73	0,61	0,43	0,35	0,24	0,07
CaMg-87-88	-0,58	0,78	0,80	0,71	0,56	0,45	0,34	0,19
CaMg-7-87/88	-0,50	0,77	0,82	0,80	0,69	0,59	0,49	0,33
CaMg-88	-0,46	0,72	0,78	0,79	0,75	0,66	0,57	0,42
CaMg-5-9-88	-0,28	0,62	0,69	0,76	0,78	0,74	0,67	0,51
CaMg-6-8-88	-0,25	0,57	0,65	0,73	0,75	0,73	0,69	0,53
CaMg-8-88	-0,31	0,56	0,65	0,70	0,67	0,65	0,61	0,47

3.1.4.4. Area-compatible values

All effect parameters have a certain area constant. Pike is a stationary predator. A value on Cs-pi represents a much larger area than the area around the catch site, because of the mobility of the prey and the prey's prey. If a lake is very large, let us say >25 km^2, and has several tributaries with different Cs-concentrations, there may be justification for dividing the lake into parts that are different in relation to each other. Our 41 lakes make up a relatively homogeneous group as regards lake size. The largest lake is 2.7 km^2 (lake 2211) and the smallest is 0.07 km^2 (lake 2214).

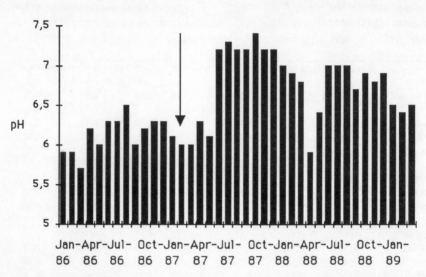

pH in Lake Selasjön, 2201; liming in March-87

Fig. 3.12. Example of the time variation within a lake for a water chemical sensitivity parameter. In this case pH in Lake Selasjön during the period January 1986 to March 1986. Lake liming was done in March 1987.

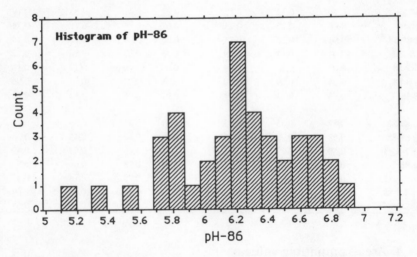

Fig. 3.13. Example of variation between lakes. In this case mean pH for 1986 in our 41 lakes.

Most lakes have, however, an area between 0.2 and 2 km^2. The mean values for the relevant effect, dose and sensitivity parameters are thus comparable in this respect.

In conclusion: The given effect, dose and sensitivity parameters have an acceptable area compatibility if they are mean values from lakes of a similar size. The best time compatibility for the dose and sensitivity parameters in relation to our effect parameters seem to be obtained for mean values from about 1 to 2 years prior to the date of the catch.

3.1.5. INTERNAL CORRELATIONS, CLUSTERS AND FUNCTIONAL GROUPS

It is clear that certain parameters are related to each other causally and methodologically, e.g., the water chemical parameters alkalinity, conductivity and hardness. There are several statistical methods of establishing internal relationships, e.g., cluster analysis or non-parametric correlations. Here, we have used simple linear correlations. The reason is that the different statistical methods often give similar results. We are mostly looking for ecological, functional relationships and the statistical methods are only an aid in our search.

3.1.5.1. Sensitivity parameters

Table 3.6 gives a correlation matrix for all "constant" environmental parameters. **Because of the definition** of these parameters, we can already at the outset say that there are certain **clusters**, e.g.:

• **Drainage area parameters** Forest% + Lake% + Mire% + Open land% originate from the same map and together form 100% of the drainage area. Thus, they belong together.

• Twenty-eight drainage areas lack basic bedrocks. Coarse sediments are not found in 13 drainage areas and fine sediments in 19 drainage areas. Together, these parameters form a group which is omitted in this work because **it is impossible to obtain the same statistical conditions for these parameters as for the others**. Naturally, this need not imply that data on these parameters cannot be used in other connections.

• Among the **lake morphometric parameters**, at least the following groups can be identified:

- Size parameters for area (i.e., total lake area and water surface) and volume (Vol).
- Form parameters which are linked to average depth, i.e., Vd (the form factor; a function of mean and max. depth) and DR (dynamic ratio; a function of lake area and mean depth).

Table 3.6. *Correlation matrix between a number of catchment area parameters, lake morphometric parameters and "process parameters" to establish clusters.*

	ADr	Br	Lake%	For%	Mire%	Ol%	Rock%	Mor%	Im%	Acid%	Lmax	Dmax	lo	Atot	area	Vol	Bm	Dm	F	Vd	DR	BET	BA	Q	T
ADr	1,00	-0,41	0,25	-0,12	0,10	-0,08	-0,14	0,06	-0,01	0,02	0,35	0,12	0,37	0,30	0,30	0,25	0,18	0,05	0,24	-0,06	0,42	0,08	-0,08	0,99	-0,27
Br	-0,41	1,00	-0,08	0,07	-0,16	0,22	0,18	-0,15	-0,14	0,17	-0,33	-0,08	-0,37	-0,33	-0,33	-0,25	-0,32	-0,09	-0,15	-0,14	-0,38	-0,24	0,24	-0,42	0,00
Lake%	0,25	-0,08	1,00	-0,38	0,15	-0,10	-0,01	-0,11	-0,05	0,04	0,06	-0,15	0,06	-0,08	-0,08	-0,10	-0,17	-0,25	0,21	0,01	0,37	-0,05	-0,21	0,22	-0,20
For%	-0,12	0,07	-0,38	1,00	-0,86	-0,17	0,19	0,26	-0,20	0,13	-0,33	0,00	-0,40	-0,42	-0,42	-0,26	-0,48	0,07	-0,13	0,03	-0,37	0,01	0,05	-0,15	-0,13
Mire%	0,10	-0,16	0,15	-0,86	1,00	-0,30	-0,06	-0,06	0,19	-0,13	0,40	0,13	0,49	0,53	0,53	0,37	0,57	0,06	0,23	-0,13	0,31	0,01	-0,01	0,14	0,25
Ol%	-0,08	0,22	-0,10	-0,17	-0,30	1,00	0,21	-0,37	0,07	-0,04	-0,15	-0,19	-0,18	-0,14	-0,14	-0,16	-0,04	-0,11	-0,29	0,21	-0,06	-0,06	0,06	-0,09	-0,12
Rock%	-0,14	0,18	-0,01	0,19	-0,06	0,21	1,00	-0,79	0,53	-0,53	-0,16	0,25	-0,21	-0,19	-0,19	-0,01	-0,24	0,40	-0,16	0,04	-0,52	-0,13	0,13	-0,15	0,15
Mor%	0,06	-0,15	-0,11	0,26	-0,06	-0,37	-0,79	1,00	-0,48	0,42	0,04	-0,15	0,04	0,04	0,04	-0,04	0,04	-0,27	0,05	-0,08	0,29	0,10	-0,10	0,06	-0,10
Im%	-0,01	-0,14	-0,05	-0,20	0,19	0,07	0,53	-0,48	1,00	-0,98	-0,18	0,39	0,23	0,21	0,21	0,32	0,07	0,43	0,18	-0,02	-0,22	0,01	-0,01	0,03	0,29
Acid%	0,02	0,17	0,04	0,13	-0,13	-0,04	-0,53	0,42	-0,98	1,00	-0,18	-0,37	-0,20	-0,18	-0,18	-0,29	-0,04	-0,43	-0,16	-0,01	0,24	0,01	-0,01	-0,01	-0,27
Lmax	0,35	-0,33	0,06	-0,33	0,40	-0,15	-0,16	0,04	-0,18	-0,18	1,00	0,57	0,95	0,89	0,89	0,87	0,42	0,45	0,49	-0,34	0,16	-0,22	0,22	0,41	0,48
Dmax	0,12	-0,08	-0,15	0,00	0,13	-0,19	0,25	-0,15	0,39	-0,37	0,57	1,00	0,56	0,54	0,55	0,75	0,36	0,90	0,16	-0,63	-0,49	-0,44	0,44	0,17	0,53
lo	0,37	-0,37	0,06	-0,40	0,49	-0,18	-0,21	0,04	0,23	-0,20	0,95	0,56	1,00	0,90	0,90	0,87	0,52	0,43	0,60	-0,31	0,22	-0,16	0,16	0,43	0,44
Atot	0,30	-0,33	-0,08	-0,42	0,53	-0,14	-0,19	0,04	0,21	-0,18	0,89	0,54	0,90	1,00	1,00	0,90	0,75	0,45	0,25	-0,32	0,21	-0,27	0,27	0,37	0,59
area	0,30	-0,33	-0,08	-0,42	0,53	-0,14	-0,19	0,04	0,21	-0,18	0,89	0,55	0,90	1,00	1,00	0,91	0,75	0,45	0,25	-0,32	0,20	-0,27	0,27	0,37	0,59
Vol	0,25	-0,25	-0,10	-0,26	0,37	-0,16	-0,01	-0,04	0,32	-0,29	0,87	0,75	0,87	0,90	0,91	1,00	0,58	0,70	0,31	-0,36	-0,05	-0,34	0,34	0,32	0,68
Bm	0,18	-0,32	-0,17	-0,48	0,57	-0,04	-0,24	0,04	0,07	-0,04	0,42	0,36	0,52	0,75	0,75	0,58	1,00	0,32	-0,18	-0,19	0,26	-0,18	0,18	0,22	0,50
Dm	0,05	-0,09	-0,25	0,07	0,06	-0,11	0,40	-0,27	0,43	-0,43	0,45	0,90	0,43	0,45	0,45	0,70	0,32	1,00	0,05	-0,33	-0,57	-0,31	0,31	0,09	0,56
F	0,24	-0,15	0,21	-0,13	0,23	-0,29	-0,16	0,05	0,18	-0,16	0,49	0,16	0,60	0,25	0,25	0,31	-0,18	0,05	1,00	-0,15	0,13	0,09	-0,09	0,26	-0,05
Vd	-0,06	-0,14	0,01	0,03	-0,13	0,21	0,04	-0,08	-0,02	-0,01	-0,34	-0,63	-0,31	-0,32	-0,32	-0,36	-0,19	-0,33	-0,15	1,00	0,40	0,59	-0,59	-0,08	-0,22
DR	0,42	-0,38	0,37	-0,37	0,31	-0,06	-0,52	0,29	-0,22	0,24	0,16	-0,49	0,22	0,21	0,20	-0,05	0,26	-0,57	0,13	0,40	1,00	0,52	-0,52	0,41	-0,19
BET	0,08	-0,24	-0,05	0,01	0,01	-0,06	-0,13	0,10	0,01	0,01	-0,22	-0,44	-0,16	-0,27	-0,27	-0,34	-0,18	-0,31	0,09	0,59	0,52	1,00	-1,00	0,06	-0,30
BA	-0,08	0,24	-0,21	0,05	-0,01	0,06	0,13	-0,10	-0,01	-0,01	0,22	0,44	0,16	0,27	0,27	0,34	0,18	0,31	-0,09	-0,59	-0,52	-1,00	1,00	-0,06	0,30
Q	0,99	-0,42	0,22	-0,15	0,14	-0,09	-0,15	0,06	0,03	-0,01	0,41	0,17	0,43	0,37	0,37	0,32	0,22	0,09	0,26	-0,08	0,41	0,06	-0,06	1,00	-0,22
T	-0,27	0,00	-0,20	-0,13	0,25	-0,12	0,15	-0,10	0,29	-0,27	0,48	0,53	0,44	0,59	0,59	0,68	0,50	0,56	-0,05	-0,22	-0,19	-0,30	0,30	-0,22	1,00

- Form parameters associated with area, i.e., Dm (mean depth), DR, Bm (mean width) and F (shore irregularity; a function of lake area and shore-line length).

• **Process parameters** such as:

(1) BA (percentage area of so-called accumulation bottoms; i.e., areas whe-re particles finer than medium silt are being continuously deposited; see Håkanson and Jansson, 1983) and BET (=100 - BA; the proportion of ero-sion and transportation bottoms; i.e. bottom where particles finer than medium silt are being discontinuously deposited). BA (and BET) are deter-mined from the dynamic ratio (DR) and are thus associated with lake area and mean depth, and
(2) the theoretical water turnover time (T), which is defined from the ratio between the lake volume (Vol) and the water transport into the lake (Q). Q is determined here from the specific runoff and the area of the drainage area (ADr).

It should be noted that there are no special connections between BA, T and Q other than that they are parameters which are "constant" according to current definitions and that they are linked with hydrodynamic and sedimentological processes. The water discharge in rivers and streams naturally varies with precipitation and time of year but here we do not have data on these variations. Here, the true frequency distribution is rep-resented by a "constant" (Q), which is intended to reflect a mean water flow. From Table 3.6, it can be seen that parameters associated definition-wise also show high correlation. Table 3.7 gives a ranking of their inter-correlation, where the criteria for this is given by the linear correlation coefficient (r). From this **r-ranking table** we can, e.g., see:

- that Q is the individual parameter that best co-varies with the size of the drainage area (ADr). This is logical and depends on Q being defined from ADr.

- that the relief (Br, i.e., a function of the maximum height difference in the drainage area; see Håkanson et al., 1990d) is not associated with any particular factor.

- that the lake percentage (Lake%) does not co-vary with any particular factor.

- that the percentage of morainic soils (Mor%) shows a clear negative association to areas with flat rocks (Rock%); the higher the Mor% the lower the Rock%.

Table 3.7. r-ranking table for a number of catchment area parameters, lake morphometric parameters and "process parameters" to establish intercorrelations.

ADr	r	Br	r	Lake%	r	Mor%	r	area	r	Dm	r	F	r	BA	r	T	r
ADr	1,00	Br	1,00	Lake%	1,00	Mor%	1,00	area	1,00	Dm	1,00	F	1,00	BA	1,00	T	1,00
Q	0,99	BA	0,24	DR	0,37	Acid%	0,42	Atot	1,00	Dmax	0,90	lo	0,60	Dmax	0,44	Vol	0,68
DR	0,42	Ol%	0,22	ADr	0,25	DR	0,29	Vol	0,91	Vol	0,70	Lmax	0,49	Vol	0,34	area	0,59
lo	0,37	Rock%	0,18	Q	0,22	For%	0,26	lo	0,90	T	0,56	Vol	0,31	Dm	0,31	Atot	0,59
Lmax	0,35	Acid%	0,17	BET	0,21	BET	0,10	Lmax	0,89	area	0,45	Q	0,26	T	0,30	Dm	0,56
Atot	0,30	For%	0,07	F	0,21	ADr	0,06	Bm	0,75	Lmax	0,45	area	0,25	Atot	0,27	Dmax	0,53
area	0,30	T	0,00	Mire%	0,15	Q	0,06	T	0,59	Atot	0,45	Atot	0,25	area	0,27	Bm	0,50
Vol	0,25	Lake%	-0,08	Lmax	0,06	F	0,05	Dmax	0,55	lo	0,43	ADr	0,24	Br	0,24	Lmax	0,48
Lake%	0,25	Dmax	-0,08	lo	0,06	lo	0,04	Mire%	0,53	Im%	0,43	Mire%	0,23	Lmax	0,22	lo	0,44
F	0,24	Dm	-0,09	Acid%	0,04	Lmax	0,04	Dm	0,45	Rock%	0,40	Lake%	0,21	Bm	0,18	BA	0,30
Bm	0,18	Vd	-0,14	Vd	0,01	Atot	0,04	Q	0,37	Bm	0,32	Im%	0,18	lo	0,16	Im%	0,29
Dmax	0,12	Im%	-0,14	Rock%	-0,01	area	0,04	ADr	0,30	BA	0,31	Dmax	0,16	Rock%	0,13	Mire%	0,25
Mire%	0,10	F	-0,15	Im%	-0,05	Vol	-0,04	BA	0,27	Q	0,09	DR	0,13	Ol%	0,06	Rock%	0,15
BET	0,08	Mor%	-0,15	Atot	-0,08	Bm	-0,06	F	0,25	For%	0,07	BET	0,09	For%	0,05	Br	0,00
Mor%	0,06	Mire%	-0,16	Br	-0,08	Mire%	-0,08	Im%	0,21	Mire%	0,06	Dm	0,05	Im%	-0,01	F	-0,05
Dm	0,05	BET	-0,24	area	-0,08	T	-0,10	DR	0,20	ADr	0,05	Mor%	0,05	Acid%	-0,01	Mor%	-0,10
Acid%	0,02	Vol	-0,25	Ol%	-0,10	BA	-0,10	Mor%	0,04	F	0,05	T	-0,05	Mire%	-0,01	Ol%	-0,12
Im%	-0,01	Bm	-0,25	Vol	-0,10	Lake%	-0,11	Lake%	-0,08	Br	-0,09	BA	-0,09	Q	-0,06	For%	-0,13
Vd	-0,06	Lmax	-0,32	Mor%	-0,11	Br	-0,15	Ol%	-0,14	Ol%	-0,11	For%	-0,13	ADr	-0,08	DR	-0,19
BA	-0,08	Atot	-0,33	Dmax	-0,15	Im%	-0,15	Acid%	-0,18	Lake%	-0,25	Br	-0,15	F	-0,09	Lake%	-0,20
Ol%	-0,08	area	-0,33	Bm	-0,17	Dm	-0,27	Rock%	-0,19	Mor%	-0,27	Vd	-0,15	Mor%	-0,10	Vd	-0,22
For%	-0,12	lo	-0,37	T	-0,20	Dmax	-0,37	BET	-0,27	BET	-0,31	Acid%	-0,16	Lake%	-0,21	Q	-0,22
Rock%	-0,14	DR	-0,38	BA	-0,21	Vd	-0,48	Vd	-0,32	Vd	-0,33	Rock%	-0,16	DR	-0,52	ADr	-0,27
T	-0,27	ADr	-0,41	Dm	-0,25	Ol%	-0,79	Br	-0,33	Acid%	-0,43	Bm	-0,18	Vd	-0,59	Acid%	-0,27
Br	-0,41	Q	-0,42	For%	-0,38			For%	-0,42	DR	-0,57	Ol%	-0,29	BET	-1,00	BET	-0,30

- that the morphometric size parameters lake area, total area (Atot), volume (Vol), shoreline length (lo) and maximum length (Lmax) are associated (r=0.89).

- that mean depth (Dm), maximum depth (Dmax), volume (Vol), theoretical water turnover time (T) and dynamic ratio (DR) form a cluster.

- that the shoreline irregularity (F) is only strongly associated (r>0.5) to the shoreline length (lo), which is self-evident from the definition of F (F=lo/2$\pi\sqrt{a}$).

- that the area of accumulation bottoms (BA) obviously has an r-value minus 1 in relation to the area of erosion plus transportation bottoms (BET) and, in other respects, clear associations to the dynamic ratio (DR) and the volume development (Vd).

- that the theoretical water turnover time (T) is markedly linked to volume, area, mean depth and maximum depth — the larger the T, the larger the lake.

> It is important to realize how the parameters group together, i.e., what functional and/or method-dependent clusters are present. For a correct step-wise multiple regression analysis, one should have a minimum of intercorrelated parameters and base final models on parameters which are functionally different.

A correlation matrix for water chemical parameters is given in Table 3.8. From this table, we can, e.g., see:

- that there is no significant correlation between pH-86 and pH-87 (r=0.07), which depends mainly on the liming done during 1987. However, there is a noticeable relationship between pH-86 and pH-88 (r= 0.74), which indicates that the conditions start to return to normal in 1988 following the intensive liming inputs made in 1987.

- that this pattern between years can be found for all parameters influenced by liming, i.e., mainly alkalinity, conductivity, hardness (CaMg), and contents of calcium and potassium. Table 3.9 shows the co-variations of the water chemical parameters before the remedial measures started. From this table and from Table 3.8, we can see:

- that the colour (Col) of the lake water is particularly associated with the iron content (Fe). This is logical and depends on the transport of humus out of, e.g., mires.

Table 3.8. Correlation matrix between a number of water chemical variables (means for 1986, 1987 and 1988) to establish clusters.

A.	pH-86	pH-87	pH-88	alk-86	alk-87	alk-88	cond-86	cond-87	cond-88	CaMg-86	CaMg-87	CaMg-88	Ca-86	Ca-87	Ca-88
pH-86	1,00	0,07	0,74	0,83	0,01	0,40	0,67	0,51	0,57	0,75	0,58	0,41	0,76	0,00	0,22
pH-87	0,07	1,00	0,02	-0,07	1,00	-0,09	-0,14	-0,63	-0,15	-0,10	-0,23	-0,16	-0,07	0,99	-0,09
pH-88	0,74	0,02	1,00	0,76	-0,06	0,72	0,66	0,62	0,78	0,73	0,78	0,73	0,72	-0,09	0,63
alk-86	0,83	-0,07	0,76	1,00	-0,12	0,58	0,95	0,76	0,86	0,98	0,87	0,70	0,98	-0,14	0,50
alk-87	0,01	1,00	-0,06	-0,12	1,00	-0,14	-0,18	-0,68	-0,21	-0,15	-0,30	-0,21	-0,12	1,00	-0,13
alk-88	0,40	-0,09	0,72	0,58	-0,14	1,00	0,50	0,50	0,85	0,56	0,67	0,95	0,59	-0,18	0,97
cond-86	0,67	-0,14	0,66	0,95	-0,18	0,50	1,00	0,80	0,87	0,99	0,89	0,69	0,98	-0,20	0,47
cond-87	0,51	-0,63	0,62	0,76	-0,68	0,50	0,80	1,00	0,75	0,79	0,87	0,63	0,76	-0,70	0,46
cond-88	0,57	-0,15	0,78	0,86	-0,21	0,85	0,87	0,75	1,00	0,88	0,90	0,96	0,89	-0,24	0,84
CaMg-86	0,75	-0,10	0,73	0,98	-0,15	0,56	0,99	0,79	0,88	1,00	0,89	0,71	1,00	-0,17	0,51
CaMg-87	0,58	-0,23	0,78	0,87	-0,30	0,67	0,89	0,87	0,90	0,89	1,00	0,79	0,88	-0,33	0,65
CaMg-88	0,41	-0,16	0,73	0,70	-0,21	0,95	0,69	0,63	0,96	0,71	0,79	1,00	0,73	-0,24	0,96
Ca-86	0,76	-0,07	0,72	0,98	-0,12	0,59	0,98	0,76	0,89	1,00	0,88	0,73	1,00	-0,13	0,54
Ca-87	0,00	0,99	-0,09	-0,14	1,00	-0,18	-0,20	-0,70	-0,24	-0,17	-0,33	-0,24	-0,13	1,00	-0,17
Ca-88	0,22	-0,09	0,63	0,50	-0,13	0,97	0,47	0,46	0,84	0,51	0,65	0,96	0,54	-0,17	1,00
K-87	0,45	-0,48	0,66	0,71	-0,53	0,83	0,71	0,83	0,88	0,73	0,84	0,88	0,73	-0,55	0,81
Col-86	-0,76	0,21	-0,52	-0,66	0,26	-0,03	-0,65	-0,61	-0,38	-0,65	-0,56	-0,15	-0,61	0,28	0,11
Col-87	-0,57	-0,79	-0,34	-0,35	-0,75	-0,02	-0,25	0,23	-0,12	-0,29	-0,13	0,01	-0,31	-0,74	0,08
Col-88	-0,85	-0,07	-0,74	-0,74	0,01	-0,29	-0,65	-0,54	-0,50	-0,68	-0,66	-0,33	-0,67	0,03	-0,14
Fe-86	-0,36	0,51	-0,29	-0,34	0,56	-0,03	-0,37	-0,59	-0,24	-0,31	-0,44	-0,11	-0,27	0,58	0,06
Fe-87	-0,49	-0,77	-0,30	-0,26	-0,72	0,03	-0,16	0,28	-0,04	-0,18	-0,07	0,07	-0,21	-0,71	0,12
Fe-88	-0,28	0,70	-0,39	-0,29	0,75	-0,25	-0,29	-0,69	-0,32	-0,26	-0,51	-0,28	-0,23	0,78	-0,18
temp-86	0,31	0,57	0,34	0,09	0,53	0,24	-0,03	-0,28	0,10	0,06	-0,03	0,13	0,06	0,51	0,17
temp-87	0,10	-0,93	0,22	0,20	-0,96	0,28	0,23	0,74	0,30	0,24	0,40	0,31	0,20	-0,96	0,25
temp-88	0,51	0,26	0,64	0,36	0,18	0,39	0,26	0,24	0,33	0,37	0,44	0,29	0,34	0,17	0,28
Sec-87	-0,02	0,99	-0,08	-0,16	1,00	-0,17	-0,21	-0,70	-0,24	-0,19	-0,32	-0,24	-0,15	0,99	-0,16
Sec-88	0,56	-0,24	0,27	0,26	-0,30	-0,19	0,19	0,33	-0,03	0,20	0,20	-0,19	0,16	-0,30	-0,36
totP-86	0,46	-0,09	0,55	0,79	-0,11	0,52	0,84	0,68	0,80	0,84	0,77	0,67	0,84	-0,13	0,53
totP-87	0,22	-0,75	0,28	0,53	-0,75	0,30	0,63	0,83	0,57	0,59	0,56	0,48	0,56	-0,75	0,32
totP-88	0,18	-0,38	0,30	0,57	-0,38	0,20	0,72	0,67	0,57	0,66	0,60	0,45	0,64	-0,38	0,28

B.	K-87	Col-86	Col-87	Col-88	Fe-86	Fe-87	Fe-88	temp-86	temp-87	temp-88	Sec-87	Sec-88	totP-86	totP-87	totP-88
pH-86	0,45	-0,76	-0,57	-0,85	-0,36	-0,49	-0,28	0,31	0,10	0,51	-0,02	0,56	0,46	0,22	0,18
pH-87	-0,48	0,21	-0,79	-0,07	0,51	-0,77	0,70	0,57	-0,93	0,26	0,99	-0,24	-0,09	-0,75	-0,38
pH-88	0,66	-0,52	-0,34	-0,74	-0,29	-0,30	-0,39	0,34	0,22	0,64	-0,08	0,27	0,55	0,28	0,30
alk-86	0,71	-0,66	-0,35	-0,74	-0,34	-0,26	-0,29	0,09	0,20	0,36	-0,16	0,26	0,79	0,53	0,57
alk-87	-0,53	0,26	-0,75	0,01	0,56	-0,72	0,75	0,53	-0,96	0,18	1,00	-0,30	-0,11	-0,75	-0,38
alk-88	0,83	-0,03	-0,02	-0,29	-0,03	0,03	-0,25	0,24	0,28	0,39	-0,17	-0,19	0,52	0,30	0,20
cond-86	0,71	-0,65	-0,25	-0,65	-0,37	-0,16	-0,29	-0,03	0,23	0,26	-0,21	0,19	0,84	0,63	0,72
cond-87	0,83	-0,61	0,23	-0,54	-0,59	0,28	-0,69	-0,28	0,74	0,24	-0,70	0,33	0,68	0,83	0,67
cond-88	0,88	-0,38	-0,12	-0,50	-0,24	-0,04	-0,32	0,10	0,30	0,33	-0,24	-0,03	0,80	0,57	0,57
CaMg-86	0,73	-0,65	-0,29	-0,68	-0,31	-0,18	-0,26	0,06	0,24	0,37	-0,19	0,20	0,84	0,59	0,66
CaMg-87	0,84	-0,56	-0,13	-0,66	-0,44	-0,07	-0,51	-0,03	0,40	0,44	-0,32	0,20	0,77	0,56	0,60
CaMg-88	0,88	-0,15	0,01	-0,33	-0,11	0,07	-0,28	0,13	0,31	0,29	-0,24	-0,19	0,67	0,48	0,45
Ca-86	0,73	-0,61	-0,31	-0,67	-0,27	-0,21	-0,23	0,06	0,20	0,34	-0,15	0,16	0,84	0,56	0,64
Ca-87	-0,55	0,28	-0,74	0,03	0,58	-0,71	0,78	0,51	-0,96	0,17	0,99	-0,30	-0,13	-0,75	-0,38
Ca-88	0,81	0,11	0,08	-0,14	0,06	0,12	-0,18	0,17	0,25	0,28	-0,16	-0,36	0,53	0,32	0,28
K-87	1,00	-0,37	0,15	-0,43	-0,40	0,22	-0,58	0,02	0,64	0,34	-0,56	0,10	0,56	0,61	0,42
Col-86	-0,37	1,00	0,40	0,86	0,74	0,34	0,58	-0,11	-0,31	-0,35	0,27	-0,80	-0,30	-0,34	-0,27
Col-87	0,15	0,40	1,00	0,61	-0,02	0,93	-0,26	-0,64	0,65	-0,46	-0,74	-0,28	-0,05	0,53	0,29
Col-88	-0,43	0,86	0,61	1,00	0,64	0,61	0,52	-0,21	-0,10	-0,52	0,03	-0,76	-0,31	-0,10	-0,12
Fe-86	-0,40	0,74	-0,02	0,64	1,00	0,10	0,90	0,25	-0,50	0,00	0,54	-0,82	0,01	-0,35	-0,16
Fe-87	0,22	0,34	0,93	0,61	0,10	1,00	-0,16	-0,44	0,70	-0,29	-0,72	-0,36	0,08	0,59	0,29
Fe-88	-0,58	0,58	-0,26	0,52	0,90	-0,16	1,00	0,27	-0,75	-0,12	0,74	-0,69	-0,01	-0,43	-0,15
temp-86	0,02	-0,11	-0,64	-0,21	0,25	-0,44	0,27	1,00	-0,33	0,58	0,54	-0,06	-0,12	-0,47	-0,49
temp-87	0,64	-0,31	0,65	-0,10	-0,50	0,70	-0,75	-0,33	1,00	0,09	-0,96	0,29	0,17	0,70	0,31
temp-88	0,34	-0,35	-0,46	-0,52	0,00	-0,29	-0,12	0,58	0,09	1,00	0,16	0,22	0,22	-0,18	-0,20
Sec-87	-0,56	0,27	-0,74	0,03	0,54	-0,72	0,74	0,54	-0,96	0,16	1,00	-0,29	-0,14	-0,76	-0,40
Sec-88	0,10	-0,80	-0,28	-0,76	-0,82	-0,36	-0,69	-0,06	0,29	0,22	-0,29	1,00	-0,20	0,03	-0,13
totP-86	0,56	-0,30	-0,05	-0,31	0,01	0,08	-0,01	-0,12	0,17	0,22	-0,14	-0,20	1,00	0,68	0,78
totP-87	0,61	-0,34	0,53	-0,10	-0,35	0,59	-0,43	-0,47	0,70	-0,18	-0,76	0,03	0,68	1,00	0,83
totP-88	0,42	-0,27	0,29	-0,12	-0,16	0,29	-0,15	-0,49	0,31	-0,20	-0,40	-0,13	0,78	0,83	1,00

Table 3.9. r-ranking table for a number of water chemical variables (means for 1986, 1987 and 1988) to establish intercorrelations.

	CaMg-86			pH-86
Ca-86	1,00		alk-86	0,83
cond-86	0,99		Ca-86	0,76
alk-86	0,98		CaMg-86	0,75
totP-86	0,84		cond-86	0,67
pH-86	0,75		totP-86	0,46
temp-86	0,06		temp-86	0,31
Fe-86	-0,31		Fe-86	-0,36
Col-86	-0,65		Col-86	-0,76
	Ca-86			**Col-86**
CaMg-86	1,00		Fe-86	0,74
alk-86	0,98		temp-86	-0,11
cond-86	0,98		totP-86	-0,30
totP-86	0,84		Ca-86	-0,61
pH-86	0,76		cond-86	-0,65
temp-86	0,06		CaMg-86	-0,65
Fe-86	-0,27		alk-86	-0,66
Col-86	-0,61		pH-86	-0,76
	cond-86			**Fe-86**
CaMg-86	0,99		Col-86	0,74
Ca-86	0,98		temp-86	0,25
alk-86	0,95		totP-86	0,01
totP-86	0,84		Ca-86	-0,27
pH-86	0,67		CaMg-86	-0,31
temp-86	-0,03		alk-86	-0,34
Fe-86	-0,37		pH-86	-0,36
Col-86	-0,65		cond-86	-0,37
	alk-86			**temp-86**
CaMg-86	0,98		pH-86	0,31
Ca-86	0,98		Fe-86	0,25
cond-86	0,95		alk-86	0,09
pH-86	0,83		CaMg-86	0,06
totP-86	0,79		Ca-86	0,06
temp-86	0,09		cond-86	-0,03
Fe-86	-0,34		Col-86	-0,11
Col-86	-0,66		totP-86	-0,12
	totP-86			
cond-86	0,84			
CaMg-96	0,84			
Ca-86	0,84			
alk-86	0,79			
pH-86	0,46			
Fe-86	0,01			
temp-86	-0,12			
Col-86	-0,30			

- that lake temperature (temp) and Secchi disc transparency (Sec) show very high correlations (r=-0.96) in some years (e.g. 1987), but considerably lower r-values in other years (e.g., 0.22 in 1988).

- that the concentration of total phosphorus (totP), which is functionally associated with the production of algae, shows a clear relationship to the water chemical cluster parameters ($r > 0.79$ for alkalinity, Ca, CaMg and conductivity for 1986; Table 3.9).

Thus, we can establish that for the water chemical parameters there are at least the following groups:

•• pH
•• Alkalinity, conductivity, hardness, calcium and potassium
•• Colour and iron
•• Temperature
•• Secchi disc transparency
•• Total-P

3.1.5.2. Dose parameters

A correlation matrix for dose parameters (and effect parameters) is given in Table 3.10. In Table 3.11 an r-ranking table is given in the same way as earlier for the sensitivity parameters. Table 3.10 shows that all caesium parameters co-vary. One might suspect that the r-values, e.g., between Cs-pe-86, Cs-pe87 and Cs-pe88 and the primary dose (Cs-soil), would become lower year by year, but this does not yet appear to be the case; the r-values between the parameters Cs-pe, Cs-bo and Cs-soil were lower in 1986 than in 1987. This indicates, e.g., that conditions during the fallout year 1986 were special and that the conditions during the subsequent years are more comparable as regards the caesium parameters.

• The primary dose (Cs-soil) is a special dose parameter. It may be regarded as a "constant" for any given lake (to the extent that Cs-soil here is determined from a given fallout map), whereas the others are typical variables.

• Caesium in lake water (Cs-wa) can be regarded separately since it is based on several water samples during the year.

• The Cs-concentrations in materials from sediment traps placed near the surface (Cs-su) and near the bottom (Cs-bo) make up a special group for methodological reasons; these are values from the given registration period. The data material from the traps placed near the surface is parti-

Table 3.10. Correlation matrix between a number of caesium parameters (means of effect and dose parameters for 1986, 1987 and 1988).

	Cs-pi87	Cs-pi88	Cs-pe86	Cs-pe87	Cs-pe88	Cs-soil	Cs-wa87	Cswa88	Cs-su86	Cs-su87	Cs-su88	Cs-bo86	Cs-bo87	Cs-bo88
Cs-pi87	1,00	0,91	0,81	0,86	0,87	0,84	0,87	0,95	0,78	0,64	0,62	0,62	0,74	0,78
Cs-pi88	0,91	1,00	0,91	0,80	0,85	0,70	0,88	0,90	0,85	0,63	0,63	0,66	0,71	0,79
Cs-pe86	0,81	0,91	1,00	0,81	0,75	0,60	0,79	0,75	0,73	0,71	0,72	0,53	0,75	0,84
Cs-pe87	0,86	0,80	0,81	1,00	0,89	0,83	0,85	0,83	0,76	0,60	0,64	0,71	0,73	0,69
Cs-pe88	0,87	0,85	0,75	0,89	1,00	0,77	0,81	0,89	0,80	0,39	0,43	0,82	0,55	0,57
Cs-soil	0,84	0,70	0,60	0,83	0,77	1,00	0,82	0,83	0,76	0,63	0,66	0,67	0,78	0,69
Cs-wa87	0,87	0,88	0,79	0,85	0,81	0,82	1,00	0,93	0,88	0,75	0,77	0,70	0,84	0,81
Cs-wa88	0,95	0,90	0,75	0,83	0,89	0,83	0,93	1,00	0,87	0,60	0,57	0,76	0,72	0,71
Cs-su86	0,78	0,85	0,73	0,76	0,80	0,76	0,88	0,87	1,00	0,63	0,68	0,88	0,79	0,75
Cs-su87	0,64	0,63	0,71	0,60	0,39	0,63	0,75	0,60	0,63	1,00	0,93	0,31	0,96	0,91
Cs-su88	0,62	0,63	0,72	0,64	0,43	0,66	0,77	0,57	0,68	0,93	1,00	0,35	0,93	0,94
Cs-bo86	0,62	0,66	0,53	0,71	0,82	0,67	0,70	0,76	0,88	0,31	0,35	1,00	0,54	0,42
Cs-bo87	0,74	0,71	0,75	0,73	0,55	0,78	0,84	0,72	0,79	0,96	0,93	0,54	1,00	0,92
Cs-bo88	0,78	0,79	0,84	0,69	0,57	0,69	0,81	0,71	0,75	0,91	0,94	0,42	0,92	1,00

Table 3.11. r-ranking table for a number of caesium parameters (means of effect and dose parameters for 1986, 1987 and 1988) to establish inter-correlations.

	Cs-pi87		Cs-soil		Cs-wa87
Cs-pi87	1,00	Cs-soil	1,00	Cs-wa87	1,00
Cs-wa88	0,95	Cs-pi87	0,84	Cs-wa88	0,93
Cs-pi88	0,91	Cs-wa88	0,83	Cs-pi88	0,88
Cs-wa87	0,87	Cs-pe87	0,83	Cs-su86	0,88
Cs-pe88	0,87	Cs-wa87	0,82	Cs-pi87	0,87
Cs-pe87	0,86	Cs-bo87	0,78	Cs-pe87	0,85
Cs-soil	0,84	Cs-pe88	0,77	Cs-bo87	0,84
Cs-pe86	0,81	Cs-su86	0,76	Cs-soil	0,82
Cs-su86	0,78	Cs-pi88	0,70	Cs-bo88	0,81
Cs-bo88	0,78	Cs-bo88	0,69	Cs-pe88	0,81
Cs-bo87	0,74	Cs-bo86	0,67	Cs-pe86	0,79
Cs-su87	0,64	Cs-su88	0,66	Cs-su88	0,77
Cs-bo86	0,62	Cs-su87	0,63	Cs-su87	0,75
Cs-su88	0,62	Cs-pe86	0,60	Cs-bo86	0,70

cularly difficult to interpret since there has been a lot of periphyton in these traps in many lakes. This means that the amount of material and the composition of the material do not reflect the deposition so well. Consequently, the results from the sediment traps placed near the surface are not discussed here.

3.1.6. TRANSFORMATIONS AND CORRELATIONS

Figure 3.14 illustrates how absolute and logarithmic values for Cs-pi87 and Cs-pi88 are distributed. The histogrammes for Cs-pi show that the values were higher during 1988 than during 1987 and also that the two distributions are skewed to the left. The two distributions of the logarithmic values will then be more uniform, as shown by the figure.

Figure 3.15 illustrates how the dose parameters Cs-soil and Cs-wa87 are distributed. It may be noted that all these frequency distributions are more or less asymmetric.

The lake conductivity has been chosen to illustrate how a sensitivity parameter is distributed in our data material. Figure 3.16 illustrates a number of frequency distributions for conductivity. The figure shows the main types of transformations that can be used in this connection:

⇨ e**cond**; this transformation will maximize the influence of high values in regressions;

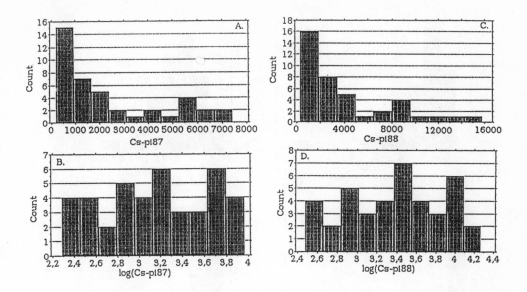

Fig. 3.14. Frequency distributions for (A) Cs-pi87 (in Bq/kg ww), (B) logarithmic values of Cs-pi87, (C) Cs-pi88 (in Bq/kg ww) and (D) log(Cs-pi88).

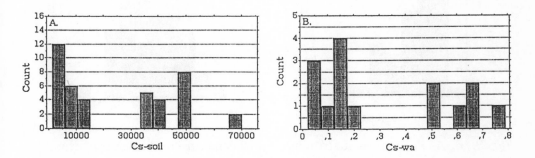

Fig. 3.15. Frequency distributions for dose parameters, (A) Cs-soil (in Bq/m^2), i.e., the primary dose (=fallout) and (B) Cs-wa (in Bq/l), i.e., the mean value of the caesium concentration in lake water for the period March 1986 - February 1987.

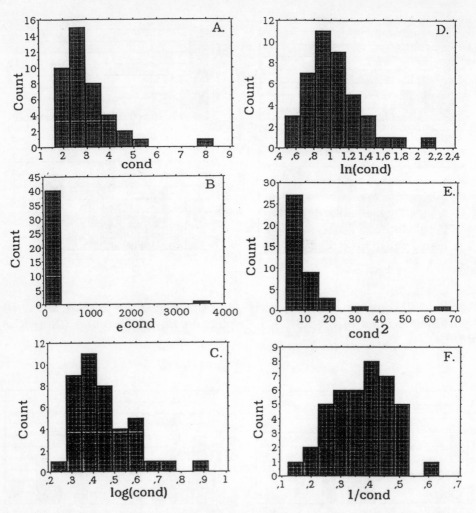

Fig. 3.16. A. Frequency distribution for the mean conductivity of the 41 lakes (for the period March 86 - February 87), and B-F five other frequency distributions for different transformations of the conductivity.

⇨ **log(cond)**, i.e., 10-logarithm for cond [or ln(cond)=the natural logarithm]; this transformation will minimize the influence of high values in regressions;

⇨ **cond^x** (where x, e.g., may be=2, or 0.5, which gives √cond, or -1, which gives 1/cond).

The intention here is to illustrate how a given material can be manipulated using a standard battery of mathematical transformations in different approaches, e.g.:

- **linear** (y=a*x + b) or

- **non-linear**, e.g., y=a*logx + b.

By transformations of this kind one can influence the degree of explanation, as illustrated by the figures in Table 3.12, where this is exemplified for Cs-pi87 and for logarithmic values of Cs-pi87 with respect to a number of caesium parameters, to different variants of conductivity and to two different adaptations, a linear adaptation and a polynomial adaptation of the second order.

If we compare the frequency distributions with the results in Table 3.12, one can clearly see that the highest r^2-values are not always obtained for the distributions that are most similar. The distribution 1/cond (Fig. 3.16) is, e.g., more similar to the distribution for log (Cs-pi87) than the distribution for Cs-pi87 (Fig. 3.14), but the r^2-value is slightly higher with regard to Cs-pi87 (0.26 as opposed to 0.21). The r^2-values obtained also, of course, apart from the distribution, depend on how well the individual pair of parameters co-vary. **Nonetheless, the frequency distribution is by no means unimportant and generally one obtains higher r^2-values when the distributions are similar.** There may also be considerable differences in this respect; from r^2=0.02 (r=-0.13 for e^{cond}) to r^2=0.26 (r=-0.51 for 1/cond; see Table 3.12).

The degree of explanation can also be influenced by the adaptation selected. This is illustrated in Table 3.12 for a normal linear adaptation (giving r^2=0.26), and for a polynomial adaptation of second order (which gives r^2=0.29). Subsequently, we will use different forms of transformations and adaptations not only for simple regressions between two variables but also for multiple regressions between several variables.

Table 3.13 is a correlation matrix for effect, dose and sensitivity parameters in relation to the effect parameter, Cs-pi87 (which is **placed in the left, upper corner**). Apart from the Cs-parameters already defined, Table 3.13 also includes a standardized effect parameter which is given by the quotient between Cs-pi and Cs-soil. This ratio is called Cs-qu87 and Cs-qu88 for pike material from 1987 and 1988, respectively. The correlation

Table 3.12. Relationship (r²-values) between Cs-pi87 and log (Cs-pi87) and different dose parameters, different parameters for caesium in perch fry and different variants of conductivity. n = 41.

	Cs-pi87	log(Cs-pi87)
Cs-soil	0.70	0.68
Cs-wa	0.76	0.75
Cs-su86	0.61	0.62
Cs-su87	0.49	0.57
Cs-bo86	0.39	0.36
Cs-bo87	0.60	0.62
Cs-pe86	0.66	0.68
Cs-pe87	0.74	0.78
Cs-pe88	0.76	0.72
Cs-pi88	0.83	0.80
cond	0.14	0.12
$_e$cond	0.02	0.01
log(cond)	0.21	0.17
ln(cond)	0.21	0.17
cond^2	0.08	0.06
$_2$cond	0.03	0.01
1/cond (linear)	0.26	0.21
1/cond (polynom)	0.29	0.22

matrix is divided into three parts. Part A links the Cs-parameters to each other, part B links Cs-pi87 to different water chemical parameters and different water chemical parameters to each other, and part C links Cs-pi87 to different morphometric and hydrological parameters where particular interest is focussed on different transformations of the water turnover time (T), to illustrate that such transformations can also be included in this connection. From Table 3.13 we can note:

• The very high correlation between Cs-pi87 and Cs-soil (r=0.84) was reduced in 1988 to r=0.70. Figure 3.17 illustrates the relationship between Cs-pi87 and Cs-soil. This figure also includes the regression line. It is probable that the impact of the fallout on the effect parameters will decrease with time after Chernobyl. In due course, this should also lead to decreasing r²-values between Cs-soil and Cs-pi.

Table 3.13. Correlation matrix in three parts: 1. Cs-pi87 vs caesium parameters, 2. Cs-pi87 vs water chemical parameters and 3. Cs-pi87 vs "constant" environmental parameters. See Appendix 2 for explanation of symbols. n = 15.

1.

	Cs-pi87	Cs-pi88	Cs-pe86	Cs-pe87	Cs-pe88	Cs-qu87	Cs-qu88	Cs-soil	Cs-su86	Cs-bo86	Cs-su87	Cs-bo87	Cs-wa87
Cs-pi87	1,00	0,91	0,81	0,86	0,87	0,00	-0,12	0,84	0,78	0,62	0,64	0,74	0,87
Cs-pi88	0,91	1,00	0,91	0,80	0,85	0,07	0,08	0,70	0,85	0,66	0,63	0,71	0,88
Cs-pe86	0,81	0,91	1,00	0,81	0,75	0,07	0,08	0,60	0,73	0,53	0,71	0,75	0,79
Cs-pe87	0,86	0,80	0,81	1,00	0,89	-0,16	-0,25	0,83	0,76	0,71	0,60	0,73	0,85
Cs-pe88	0,87	0,85	0,75	0,89	1,00	-0,01	-0,09	0,77	0,80	0,82	0,39	0,55	0,81
Cs-qu87	0,00	0,07	0,07	-0,16	-0,01	1,00	0,87	-0,44	-0,10	-0,13	-0,17	-0,25	-0,11
Cs-qu88	-0,12	0,08	0,08	-0,25	-0,09	0,87	1,00	-0,56	-0,11	-0,17	-0,17	-0,30	-0,16
Cs-soil	0,84	0,70	0,60	0,83	0,77	-0,44	-0,56	1,00	0,76	0,67	0,63	0,78	0,82
Cs-su86	0,78	0,85	0,73	0,76	0,80	-0,10	-0,11	0,76	1,00	0,88	0,63	0,79	0,88
Cs-bo86	0,62	0,66	0,53	0,71	0,82	-0,13	-0,17	0,67	0,88	1,00	0,31	0,54	0,70
Cs-su87	0,64	0,63	0,71	0,60	0,39	-0,17	-0,17	0,63	0,63	0,31	1,00	0,96	0,75
Cs-bo87	0,74	0,71	0,75	0,73	0,55	-0,25	-0,30	0,78	0,79	0,54	0,96	1,00	0,84
Cs-wa87	0,87	0,88	0,79	0,85	0,81	-0,11	-0,16	0,82	0,88	0,70	0,75	0,84	1,00

2.

	Cs-pi87	Tr-86	pH-86	alk-86	cond-86	Col-86	totP-86	Fe-86	Ca-86	CaMg-86	K-86*
Cs-pi87	1,00	-0,54	-0,51	-0,36	-0,26	0,08	-0,36	-0,22	-0,38	-0,34	-0,30
Tr-86	-0,54	1,00	0,22	-0,02	-0,14	-0,16	-0,12	0,13	-0,07	-0,08	-0,02
pH-86	-0,51	0,22	1,00	0,81	0,68	-0,71	0,44	-0,33	0,76	0,75	0,59
alk-86	-0,36	-0,02	0,81	1,00	0,95	-0,62	0,79	-0,30	0,96	0,97	0,91
cond-86	-0,26	-0,14	0,68	0,95	1,00	-0,58	0,80	-0,31	0,98	0,99	0,95
Col-86	0,08	-0,16	-0,71	-0,62	-0,58	1,00	-0,33	0,75	-0,53	-0,57	-0,60
totP-86	-0,36	-0,12	0,44	0,79	0,80	-0,33	1,00	-0,01	0,78	0,80	0,87
Fe-86	-0,22	0,13	-0,33	-0,30	-0,31	0,75	-0,01	1,00	-0,21	-0,25	-0,34
Ca-86	-0,38	-0,07	0,76	0,96	0,98	-0,53	0,78	-0,21	1,00	0,99	0,91
CaMg-86	-0,34	-0,08	0,75	0,97	0,99	-0,57	0,80	-0,25	0,99	1,00	0,93
K-86*	-0,30	-0,02	0,59	0,91	0,95	-0,60	0,87	-0,34	0,91	0,93	1,00

3.

	Cs-pi87	area	Vol.	Dm	BA	Q	T	ADr	Br	Lake%	1/T	√T	T^2	log(T)	10^T
Cs-pi87	1,00	0,14	0,08	0,38	0,49	0,58	0,00	0,57	0,00	0,02	0,60	0,17	0,15	0,38	0,22
area	0,14	1,00	0,93	0,52	0,23	0,17	0,34	0,14	0,32	0,01	0,37	0,40	0,27	0,43	0,28
Vol.	0,08	0,93	1,00	0,77	0,07	0,29	0,30	0,27	0,30	0,04	0,32	0,36	0,23	0,40	0,25
Dm	0,38	0,52	0,77	1,00	0,61	0,33	0,12	0,33	0,25	0,06	0,23	0,18	0,08	0,25	0,10
BA	0,49	0,23	0,07	0,61	1,00	0,25	0,10	0,26	0,09	0,01	0,10	0,11	0,08	0,10	0,08
Q	0,58	0,17	0,29	0,33	0,25	1,00	0,49	1,00	0,43	0,01	0,73	0,60	0,37	0,72	0,32
T	0,00	0,34	0,30	0,12	0,10	0,49	1,00	0,51	0,15	0,13	0,41	0,96	0,96	0,80	0,92
ADr	0,57	0,14	0,27	0,33	0,26	1,00	0,51	1,00	0,43	0,02	0,73	0,63	0,39	0,74	0,35
Br	0,00	0,32	0,30	0,25	0,09	0,43	0,15	0,43	1,00	0,36	0,18	0,19	0,11	0,21	0,11
Lake%	0,02	0,01	0,04	0,06	0,01	0,01	0,13	0,02	0,36	1,00	0,10	0,09	0,15	0,01	0,18
1/T	0,60	0,37	0,32	0,23	0,10	0,73	0,41	0,73	0,18	0,10	1,00	0,61	0,24	0,86	0,20
√T	0,17	0,40	0,36	0,18	0,11	0,60	0,96	0,63	0,19	0,09	0,61	1,00	0,86	0,93	0,81
T^2	0,15	0,27	0,23	0,08	0,08	0,37	0,96	0,39	0,11	0,15	0,24	0,86	1,00	0,63	0,99
log(T)	0,38	0,43	0,40	0,25	0,10	0,72	0,80	0,74	0,21	0,01	0,86	0,93	0,63	1,00	0,58
10^T	0,22	0,28	0,25	0,10	0,08	0,32	0,92	0,35	0,11	0,18	0,20	0,81	0,99	0,58	1,00

• The Cs-concentrations in lake water show even stronger relationships with Cs-pi87 than that shown by Cs-soil (r=0.87 for 1987 and 0.88 for 1988). The regression for 1987 is given in Fig. 3.17B.

• The relationship between Cs-pi87 and caesium in sediment traps is given in Fig. 3.17. Since much caesium is bound to suspended particles which settle and are resuspended in lakes, the data on Cs-concentrations in sediment traps provide an indirect measure of the total load of caesium in the lakes. But there are problems with regard to the interpretation of these values: How much of the material deposited in the traps is primary and how much is secondary (i.e., resuspended)? How old is the material? How much is allochthonous and autochthonous? It may be difficult to give scientifically relevant answers to these questions (see Håkanson et al., 1989). Nonetheless, we can note that considerable advantages are obtained by using Cs-concentrations in material from sediment traps as a measure of the lake dose in these ecosystem contexts. Not only do we get high correlations with the different effect parameters and with other dose parameters but we also get high Cs-activities in the sediment trap material, which is thus easy and relatively inexpensive to collect and analyse, and only a few analyses are required; one or two samples from sediment traps, where the natural processes ensure that the samples are integrated in time and space, providing information in a way that can otherwise only be achieved with relatively many and expensive water analyses.

• The water chemical cluster parameters (from 1986) co-vary significantly with Cs-pi87.

• Among the lake parameters, particularly the lake volume, lake area, mean depth and the theoretical water turnover time show marked correlations with Cs-pi87, i.e., the faster the turnover time (the lower the T), the lower the Cs-concentration in pike and vice versa.

• The different transformations of T do not increase the correlation to Cs-pi87; the simplest variant gives the highest r^2. In ecometric contexts, it is **a golden rule to use simple and logical relationships.**

• The different ratios (Cs-pi/Cs-soil) do not appear to provide any additional information in this context.

• There are very marked relationships between the Cs-concentrations in pike and those in perch (r=0.81 to 0.91).

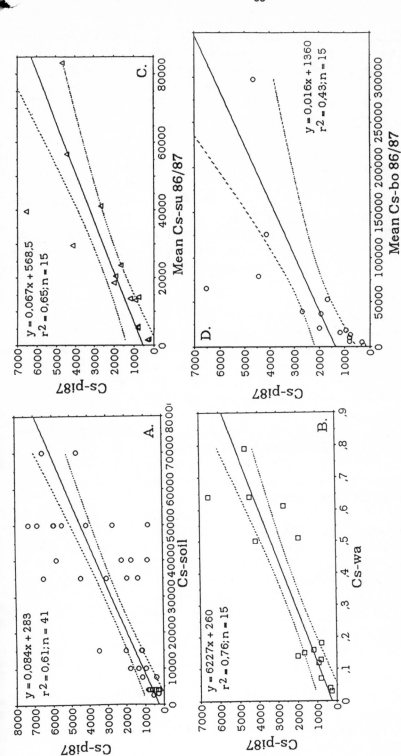

Fig. 3.17. A. Relationship between Cs-pi87 (Bq/kg ww) and Cs-soil (Bq/m²). The figure gives the regression line, its equation and the 90% confidence interval (dashed lines) as well as the r²-value.
B. Relationship between Cs-pi87 and Cs-wa (Bq/l), i.e., the mean value for caesium concentration in lake water from March 86 - February 87.
C. Relationship between Cs-pi87 and the mean value of the caesium concentration in material from sediment traps placed close to the surface for 1986 and 1987 (Cs-su; Bq/kg dry weight).
D. Relationship between Cs-pi87 and the mean value of the caesium concentration in material from sediment traps placed close to the bottom for 1986 and 1987 (Cs-bo; Bq/kg dry weight).

The high r-values to conductivity are particularly interesting because there are interesting principal relationships between conductivity and the effect parameters. The links to the potassium concentration in water (K-86* stands for the mean value for the period March 86 to Feb. 87) are also interesting, since there may also be **causal** links between the uptake of Cs and K in fish.

3.1.6.1. Links to conductivity

The lower the conductivity, the fewer the ions such as Ca, Mg and K and the larger the caesium uptake in fish. Since the highest r^2-values (according to Table 3.12) were obtained for 1/cond, we shall use this as our startingpoint. Figure 3.18A shows the best regression which is based on a least square adaptation to a second order polynom. The degree of explanation is 29% (p<0.001). The equation of the regression line is also given in the figure (i.e., $y = 1826 - 9712x + 28016x^2$). The linear adaptation, which gives a slightly lower degree of explanation (26%, p>0.001) is given in Fig. 3.18B. These are two basic types of adaptations; the third basic type (where the exponent is also varied, and use is not only made of the exponent 1) is illustrated in Fig. 3.18C. In this case, the regression has been made between Cs-pi87 and $1/cond^3$. The r^2-value then increases slightly (up to 0.292). This is also the best adaptation that can be obtained with $1/cond^x$. Thus, e.g., for the exponent x=2 we get r^2=0.284, for x=1/2 we get r^2=0.237 and for x=4 we get r^2=0.286. It should be emphasized that these manipulations, e.g., polynomial adaptation instead of simple linear adaptation or adaptation with different exponents, only provides information on the least square sum; **they need not reveal anything about the real causal relationships between the parameters.**

In addition, the adaptation that gives the highest r^2-value for a given empirical material need not necessarily give the best adaptation to another empirical material. This is illustrated in Fig. 3.18D, which shows the relationships between Cs-pi**88** and 1/cond. From Fig. 3.18 we can note:

• that the relationships between Cs-pi**88** and conductivity is generally stronger than between Cs-pi**87** and conductivity,

• that the regressions have slightly different constants and intercepts, and

• that it is not the same relationship that gives the highest degree of explanation; for the 1988 pike material, we do not get, e.g., the best adaptation for $1/cond^3$, as was obtained for the 1987 material, but for 1/cond. The differences are small (not significant) but they are included here to emphasize that **one should not place excessive reliance on empirical con-**

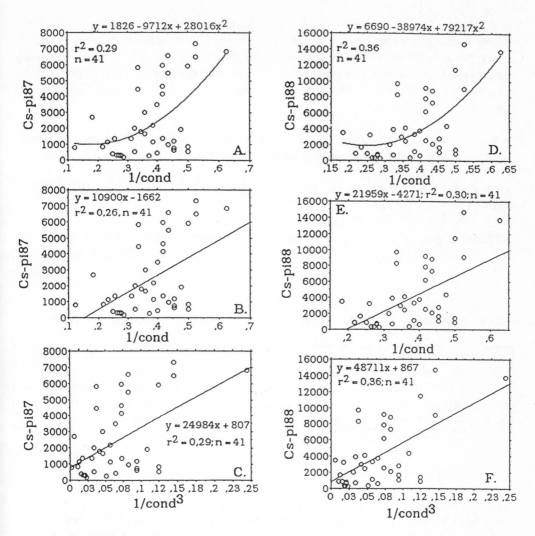

Fig. 3.18. Six different relationships between caesium concentration in pike (Bq/kg ww) and different transformations of the conductivity (in mS/m). The figure gives the regression line, its equation and the r^2-value.

A. Cs-pi87 against 1/cond in a polynomial adaptation.
B. Cs-pi87 against l/cond in a linear adaptation.
C. Cs-pi87 against l/cond³.
D. Cs-pi88 against l/cond in a second order adaptation.
E. Cs-pi88 against l/cond in a linear adaptation.
F. Cs-pi88 against l/cond³.

stants and coefficients. The important factor is, in fact, not the numerical values of the coefficients but the more overall relationships between the different parameters.

It is also a fact that the type of adaptation that gives the highest r^2-value when **two** parameters are compared pair-wise, e.g., an effect and a sensitivity parameter as in Fig. 3.18, need not give the highest r^2-value when an **additional** parameter is included, e.g., a dose parameter. This is illustrated in Fig. 3.19., where results are compared from four empirical, correlative models.

• Model 1 gives the linear relationship between Cs-pi87 (effect), Cs-soil (primary dose) and conductivity (sensitivity); the r^2-value is 0.77.

• In Model 2 we use plus (+) 1/cond instead of minus (-) cond; this gives a lower r^2-value (0.75).

• Model 3 gives Cs-pi87 as a function of the ratio between Cs-soil and cond in a linear adaptation; this gives an r^2-value of 0.80, which is a high r^2-value in this context.

• In Model 4 we give Cs-pi87 as a function of the ratio Cs-soil/cond^3. This gives an r^2-value of 0.70, which is considerably lower than in Model 3.

In summary, these examples are meant to show that there are many possibilities for combining a dose and a sensitivity parameter. The number of possible combinations with several parameters (e.g., further sensitivity parameters) increases extremely rapidly. One then has to systematically investigate which combinations give the best and most reasonable results.

3.1.6.2. Links to potassium

Since potassium and caesium can be taken up in fish in about the same way (see Black, 1957; Fleishman, 1963; Carlsson, 1978), we have considered it interesting to study the **statistical** relationships between the K-concentration in our lakes and the Cs-concentration in pike. Figure 3.20A illustrates this for the entire material from 1987 and 1988. The figure shows that extremely high Cs-pi values only occur in lakes with low K-concentrations (below 12 µeq/l). It may also be interesting to see what the relationship looks like if we standardize the caesium values in fish to the fallout. Consequently, the Cs-pi/Cs-soil ratio in Fig. 3.20B is compared with K-values for 1987 and 1988. Once again, this figure clearly shows how this ratio increases very markedly at low K-values. The results indicate that the K-value, in the same way as conductivity, is best related to Cs-pi via a reciprocal approach. Thus, Fig. 3.20C also includes the relationship between

57

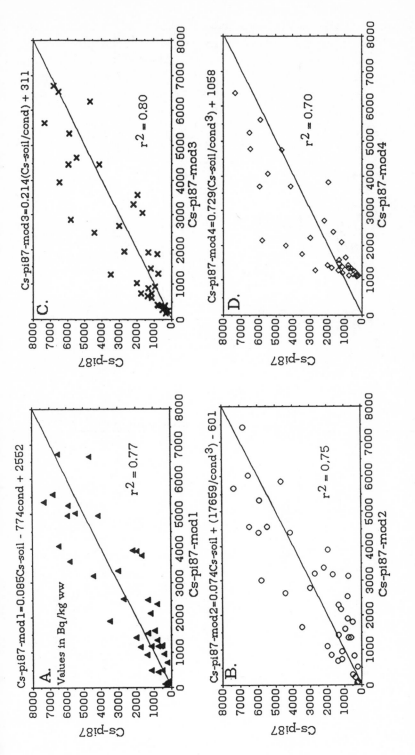

Fig. 3.19. Four different relationships between model data on caesium concentration in pike (Bq/kg ww) and empirical data from our 41 lakes. The figure gives the regression line, its equation and the r²-value.
A. Cs-pi87 against data from model 1 (which is based on Cs-soil as dose parameter and conductivity as sensitivity parameter).
B. Cs-pi87 against data from model 2 (which is based on Cs-soil and cond³).
C. Cs-pi87 against data from model 3 (which is based on the Cs-soil/cond ratio).
D. Cs-pi87 against data from model 4 (which is based on the Cs-soil/cond³ ratio).

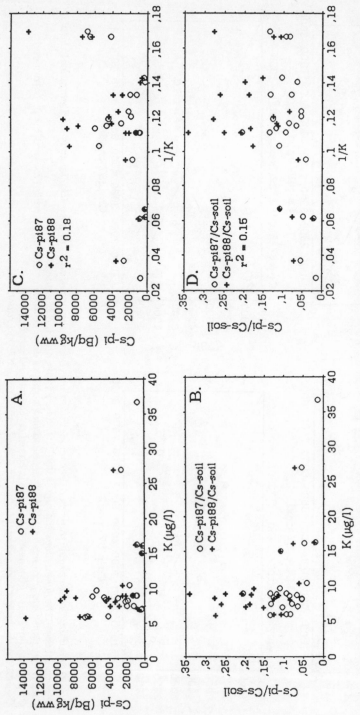

Fig. 3.20. A. Relationship between caesium concentration in pike for 1987 and 1988 and the mean concentration of potassium in the lakes (March 86 - February 87).
B. Relationship between the Cs-pi/Cs-soil ratio (for the 1987 and 1988 data; Cs-pi in Bq/kg ww; Cs-soil in Bq/m², and the K-concentration in the lakes.
C. Relationship between Cs-pi and 1/K; r²=0.18.
D. Relationship between Cs-pi/Cs-soil ratio and 1/K; r²=0.15.

Cs-pi and 1/K, as well as the relationship between the Cs-pi/Cs-soil ratio and 1/K (Fig. 3.20D).

It can be seen from Fig. 3.21 that the conductivity has higher r^2-values with regard to Cs-pi and the ratio Cs-pi/Cs-soil than those given by the K-concentration (Fig. 3.21). The most probable explanation to this may very well be linked to the simple fact that we have a rather limited data-set for K. And from this poor data-set it may not be meaningful to construct theories on causal relationships.

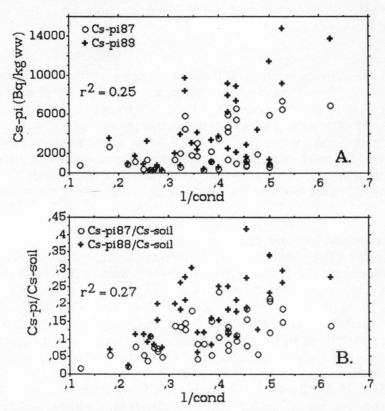

Fig. 3.21. A. Relationship between the caesium concentration in pike 1987 and the ratio 1/conductivity (mean values for March 86 - February 87; cond in mS/m); $r^2=0.25$.
B. Relationship between the Cs-pi/Cs-soil ratio (for 1987 and 1988; Cs-pi in Bq/kg ww; Cs-soil in Bq/m²) and the mean conductivity of the lakes; $r^2=0.27$.

3.2. LOAD DIAGRAMME FOR CAESIUM IN PIKE

The intention in this section has been to develop a load diagramme (effect-dose-sensitivity diagramme) for lakes that give a statistical degree of explanation as high as possible and which simultaneously is as simple and as relevant as possible. The normal approach is first to make a proper survey of what the parameters represent. We then enter area- and time-compatible data in a matrix and finally conduct step-wise multiple regression analysis. The final stage is, in fact, the easiest. An example from a step-wise multiple regression analysis is given in Table 3.14. The table is divided into three parts.

• **Part 1 — caesium parameters.**
The highest r^2-value in relation to Cs-pi87 is given by the dose parameter Cs-wa* (r^2=0.76; Cs-wa* represents the mean value for the period March 1986 to Feb. 1987). The partial correlation coefficients then show that the dose parameters Cs-su86, Cs-bo87, etc, in this order can increase the r^2-value. In the next step, we show that the r^2-value only increases from 0.76 to 0.78 when Cs-bo87, the second most important factor, is included. Consequently, the additional information obtained, if we include Cs-bo87, is very limited. This is because Cs-wa* and Cs-bo87 have a high inter-correlation.

• **Part 2 — standard parameters.**
This part of the table concerns water chemical standard parameters and Cs-soil. Altogether 17 different environmental parameters have been tested and the table shows that Cs-soil is the individual parameter that shows the highest r^2-value in relation to the effect parameter Cs-pi87 (r^2=0.59); i.e., 59% of the variation between the lakes in Cs-pi87 can be statistically explained by variations in Cs-soil. The regression line is given in the table as well as a ranking of the importance of the tested factors. It can be seen that conductivity is the second most important factor. If also this factor is also included, then the r^2-value increases from 0.59 to 0.77 (cf. Fig. 3.19A). In this regression model with two variables, we thus include conductivity as a sensitivity parameter. **Since a large number of water chemical variables show close a relationship to conductivity (alkalinity, K, Ca, CaMg), this two-variable formula also includes the influence from these cluster parameters.** Thus, after including conductivity, we get a completely different ranking of the factors. The table shows that water turnover time is now the third most important factor; water turnover time was in seventh place before we included conductivity.

In the next step, the theoretical water turnover time is accounted for. This increases the r^2-value to 0.86. The table shows how the r^2-value can increase to 0.88 if we include four variables (we then account for the mean

*Table 3.14. Result of step-wise multiple regression analysis to sta-
tistically explain the variation in Cs-pi87 between lakes. The analysis is
based on lake mean values for the period March 1986 to February 1987
from 41 lakes. The degree of explanation (r^2), the partial correlation
coefficients and the equations of the regression lines are given.*

---

1. Cs-pi87 vs Cs-parameters

	Parameter	r^2	Part. corr. coeff.
Best	Cs-wa*	0.76	
Regression line:	Cs-pi87=260 + 6227Cs-wa		
Then follows	Cs-su86		0.08
	Cs-bo87		0.07
	Cs-bo86		0.04
	Cs-su87		0.01
Best two	Cs-wa*, Cs-su86 0.780		

---

2. Cs-pi87 vs standard parameters from regular control programmes

	Parameter	r^2	Part. corr. coeff.
Best	Cs-soil	0.59	
Regression line:	Cs-pi87=135 + 0.076Cs-soil		
Then follows	cond		-0.69
	alk		-0.68
	CaMg		-0.67
	K		-0.66
	Ca		-0.65
	Water turnover time (T)		0.63
	Lake area (A)		0.62
	totP		-0.60
	Lake volume (V)		0.51
	pH		-0.45
	Mean depth (Dm)		0.18
	Dynamic ratio (DR)		0.17
	Area accumulation bottoms (BA)		0.13
	Fe		-0.05
	Water discharge (Q)		0.05
	Colour		-0.03
Best two	Cs-soil, cond	0.77	
Regression line:	Cs-pi87=2032 + 0.078Cs-soil - 661cond		
Then follows	Water turnover time		0.55
	Lake area		0.53
	Lake volume		0.40
	Colour		-0.40
	Area accumulation bottoms (BA)		0.29
	totP		-0.27
	Dynamic ratio (DR)		0.20
	Ca		0.16
	Fe		-0.13
	alk		-0.12
	pH		0.11

	CaMg	0.06
	Mean depth	-0.05
	K	-0.04
	Water discharge (Q)	-0.02

Best three	Cs-soil, cond, T give r^2=0.86	

Regression line:	Cs-pi87=1233+0.073Cs-soil-510cond+893T	

Then follows	Mean depth	-0.44
	Dynamic ratio (DR)	0.28
	CaMg	0.24
	alk	-0.22
	K	-0.21
	Fe	0.21
	Lake area	0.18
	Water discharge (Q)	0.16
	Ca	0.15
	totP	-0.10
	Lake volume	-0.08
	Area accumulation bottoms (BA)	0.06
	Colour	0.02

Best four	Cs-soil, cond, T, Dm give r^2=0.88
Best five	Cs-soil, cond, T, Dm, A give r^2=0.90

3. Cs-pi87 vs catchment parameters

	Parameter	r^2	Part. corr. coeff.
Best	Im%	0.40	

Regression line:	Cs-pi87=792 + 36.1Im

	Then follows	
Then follows	Mire%	0.38
	Ol%	-0.31
	Fs%	-0.30
	For%	-0.18
	ADr%	0.14
	Bas%	-0.12
	Acid%	0.12
	Lake%	-0.11
	Br	-0.03
	Rock%	-0.00

Best two	Im%, Mire% give r^2=0.49

Regression line:	Cs-pi87=-233+74.5Mire%-32.9Im%

Best three	Im%, Mire%, For% give r^2=0.53

Best four	Im%, Mire%, For%, Rock% give r^2=0.56

where Im=% intermediate rocks within the catchment area ; Mire%=mires; Ol%=open land; Fs%=fine sediment; For%=forest land; ADr%=Drainage area; Bas%=basic rocks; Acid%=acid rocks; Lake%=lakes; Br=relief of drainage area; Rock%=rocks, see Håkanson et al. (1988).

depth, Dm) and to 0.9 if we include five variables (then including lake area, A). But T, Dm and A are physically related to each other, **they add very little to the r^2-value and the regression models with Dm and A should not be used.**

• Part 3 — catchment parameters.
Here Cs-pi87 is linked to a number of factors describing the catchment areas of the lakes. The conditions in the catchment areas influence the limnological character of the lakes (see e.g., Knoechel and Campbell, 1988; Rochelle et al., 1989; Nilsson and Håkanson, 1991). The individual factor giving the highest r^2-value in relation to Cs-pi87 is Im%, i.e., the percentage of intermediary rock types (rock types between acidic and basic); the higher the Im%-value the higher the Cs-pi87. The percentage of mires is the next most important factor and the percentage of forests the third factor of importance in this connection. The r^2-value increases from 0.40 for Im% to 0.49 when we include Mire% and to 0.53 when we also include For%.

Figure 3.22A shows that the r^2-value for the relationship between Cs-pi87 and Cs-soil increases from 0.61 (Fig. 3.17A) to 0.73 if we convert the Cs-parameters into logarithms. The relationships that have given the highest r^2-values for two parameters (primary dose plus a sensitivity parameter) are given by Model 6 in Fig. 3.22B; i.e., the following regression relationship between Cs-pi87, Cs-soil and conductivity:

$$\log(\text{Cs-pi87}) = 0.843 \log(\text{Cs-soil}) - 0.609 \sqrt{\text{cond}^*} + 0.66 \qquad (3.4)$$
$$(r^2 = 0.88, \ n = 41)$$

where Cs-pi87=Cs-concentration in pike 1987 in Bq/kg ww; Cs-soil= primary dose determined from Fig. 2.1 in Bq/m^2; cond*=the mean conductivity of lake water during the period March 1986 to February 1987 in mS/m.

When giving a model of this kind, it is also essential to state the conditions imposed, i.e., the range shown by the parameters included in the empirical base material. If one wants to use an equation of this kind to predict conditions in other lakes, then it should only be done for lakes within the range applying to the equation/model. Thus, for Eq. (3.4) the following conditions apply:

Cs-soil	cond*
2500-70 000	1.6-8.2

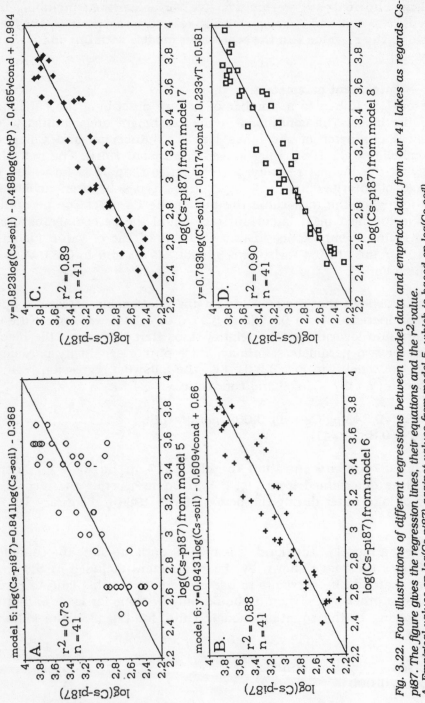

Fig. 3.22. *Four illustrations of different regressions between model data and empirical data from our 41 lakes as regards Cs-pi87. The figure gives the regression lines, their equations and the r^2-value.*
A. *Empirical values on log(Cs-pi87) against values from model 5, which is based on log(Cs-soil).*
B. *log(Cs-pi87) against values from model 6, which is based on log(Cs-soil) and √cond).*
C. *log(Cs-pi87) against values from model 7, which is based on log(Cs-soil), log(totP) (totP is given in µg/l) and √cond.*
D. *log(Cs-pi87) against values from model 8, which is based on log(Cs-soil), √cond and √T.*

Conductivity can be replaced by the water chemical cluster parameters, i.e., mainly alkalinity, CaMg and Ca, but this will then influence the r^2-value as well as the empirical constants. The r^2-value is thus higher in Eq. (3.4) than in the examples given in Fig. 3.18. This is also the best and simplest relationship between the three parameters that has been obtained (see Fig. 3.22B for a graphical illustration).

The best relationships between Cs-pi87, Cs-soil, cond and one further sensitivity parameter are given below for both totP and T, which are the parameters which yield the most in this context:

$$\log(\text{Cs-pi87})=0.823\log(\text{Cs-soil})-0.465\sqrt{\text{cond}^*}-0.488\log(\text{totP}^*)+0.984 \qquad (3.5)$$
$$(r^2=0.89, \; n=41)$$

where totP*=mean concentration of total phosphorus in lake water during the period March 1986 to February 1987 in µg/l.

For Eq. (3.5) the following conditions apply:

Cs-soil	cond*	totP*
2500-70 000	1.6-8.2	4.3-26.4

$r^2=0.89$ is a very high degree of explanation.

Models that yield higher r^2-values than 0.7 (when tested against reliable empirical data) can generally be used in practice in water management.

The relationship between model data according to Eq. (3.5) and empirical data is given in Fig. 3.22C. In an earlier section, we have shown that in practice it is impossible to reach r^2-values higher than 0.95-0.97 for Cs-pi. The formula shows that Cs-pi87 increases with the primary dose; Cs-pi87 also decreases with decreased conductivity and with decreasing totP-concentrations. This is in good agreement with earlier experience as regards caesium uptake in 1+perch (Håkanson et al., 1988) — the higher the totP-concentrations the higher the bioproduction and the greater the "biological dilution" of a given caesium dose in the lake biomass.

The other formula is given by:

$$\log(\text{Cs-pi87})=0.783\log(\text{Cs-soil})-0.517\sqrt{\text{cond}^*}+0.233\sqrt{T}+0.581 \qquad (3.6)$$
$$(r^2=0.90, \; n=41)$$

where T=the theoretical water turnover time of the lake water in years.

For Eq. (3.6) the following conditions apply:

Cs-soil	cond*	T
2500-70 000	1.6-8.2	0.02-2.9

The degree of explanation for this relationship is more than 90% and in Fig. 3.22D we can see the close fit between model and empirical data.

There are many statistical methods (from simple linear correlations and regressions, to cluster analysis, dendrogrammes and various non-para-metric tests) available to express relationships between variables, like between the empirical and the model data in Fig. 3.22. The focus in this publication is not on the benefits and restrictions linked to different statistical methods, but rather on the ecological representativity and information value of each data. So, from an ecometric perspective, it is often most interesting to view the actual scatter of the data in plots like Fig. 3.22; the different statistical methods are used as tools to **quantify** the scatter/correlation between the data. Figure 3.23 gives an alternative so-called scatter plot matrix where empirical data and models data [on log (Cs-pi87), as in Fig. 3.22] are compared directly. The plots illustrate empirical data on log(Cs-pi87) versus:

1. Absolute empirical data on Cs-pi87.

2. Model data on log(Cs-pi87); from models 5, 6, 7 and 8 (see Fig. 3.22).

3. Model data on Cs-pi87; from model 8.

The idea with Fig. 3.23 is to visualize the linkages between all the pairs of data in a matrix arrangement. The ellipses enclose about 95% of the points (for bivariate normal distributions when the number of points is large). The correlation between the model data and the empirical data is seen by the collapsing of the ellipse along the diagonal y=x axis. If the ellipse is fairly round and is not diagonally oriented, there is no correlation between the data pairs.

Another use of the scatter plot (not shown in Fig. 3.23, though) concerns the possibilities to mark, e.g., individual points, functional groups, clusters and/or lakes from certain regions to see where such points appear in the general pattern.

Since Eq. (3.6) is the best (and simplest) relationship obtained between 3 variables and the effect parameter, **we have prepared 2 "classical" 3-D load diagrammes for this model (Fig. 3.24)**. The water turnover time (T) is kept constant (1 year) in the diagramme in Fig. 3.24A, but is varied in the diagramme in Fig. 3.24B (0.02 years in geometric steps to 2.56

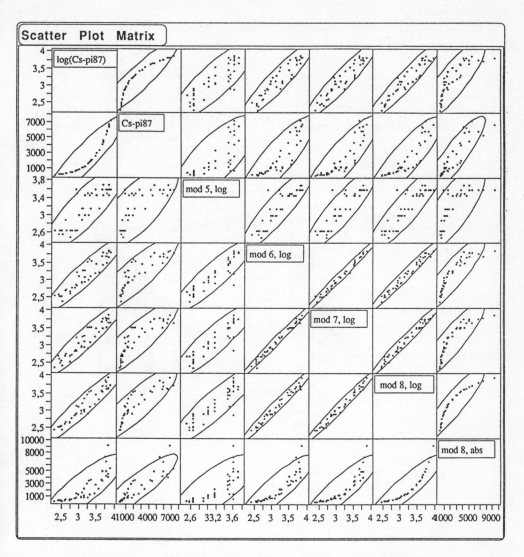

Fig. 3.23. Scatter plot matrix for empirical data of log(Cs-pi87) versus absolute empirical values of Cs-pi87, model data of log(Cs-pi87) (model 5, 6, 7 and 8), and absolute values of Cs-pi87 according to model 8; with 95% ellipses imposed on each plot.

years). The conductivity varies in steps of 1.6 mS/m in Fig. 3.24A; Cs-soil is given on the positive x-axis and the effect term on the positive y-axis. From these diagrammes, it is easily seen how the parameters are related to each other.

These results can be used in practical environmental work in several ways. For example, in an hypothetical lake (with properties within the model's

Load diagramme - Caesium in lake pike, model 8

EFFECT PARAMETERS:

- Cs-137 in fish for human consumption
- •• Cs-137 in pike; area constant=entire lake (if a<10 km^2); time constant≈2 years
- •• Cs-137 in perch fry (1+perch); area constant=entire lake; time constant≈1 year

- No known threats to the aquatic ecosystem

DOSE (=LOAD) PARAMETERS:

- Fallout of Cs-137 (=primary load)
- Secondary load (=discharge from catchment)
- Mean lake concentrations of Cs-137 as determined from:
- •• water samples
- •• sediment trap material or
- •• surficial sediment data

SENSITIVITY PARAMETERS:

- Lake water retention time
- Lake water chemistry of the "cluster" parameters, i.e., cond, alk, CaMg or pH; mean values for the entire lake for a time equal to the time constant.
- Lake morphometry, especially "cluster" parameters linked to resuspension, i.e., mean depth, dynamic ratio and BET (=area of erosion and transportation).

For T = 1 year

Cs-pi87 in Bq/kg ww
Cs-soil in Bq/m^2
Cond in mS/m

1500 Bq/kg ww -
Swedish guideline

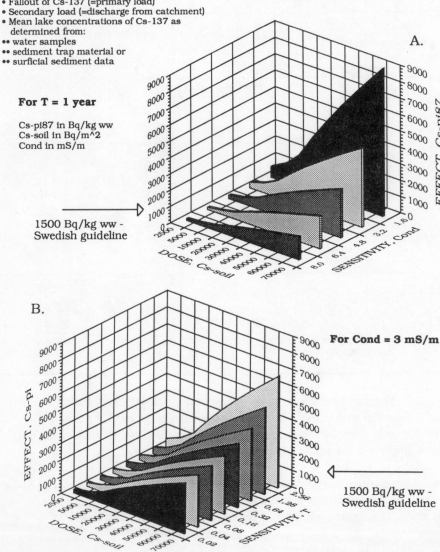

Fig. 3.24. Load diagramme for caesium in pike in 1987 according to model 8, which gives Cs-pi87 as a function of Cs-soil, conductivity and the theoretical water turnover time (T).
A. Nomogram where T=1 year.
B. Nomogram where Cond=3 mS/m.

range) with a fallout of 50 000 Bq/m^2, a conductivity of 5 mS/m and a water turnover time of 0.1 year, we can calculate a probable value for Cs-concentration in pike (for 1987) to about 1500 Bq/kg ww, which corresponds to the Swedish guideline (also marked in the Fig. 3.24).

The best model obtained (using the given conditions) gives a 91% degree of explanation. It is based on Cs-soil, conductivity, totP and T:

$$\log(\text{Cs-pi87}) = 0.781\log(\text{Cs-soil}) - 0.457\sqrt{\text{cond}^*} - 0.009\text{totP}^* + 0.2\sqrt{T} + 0.603 \qquad (3.7)$$
$$(r^2 = 0.91, \ n = 41)$$

Model conditions:

Cs-soil	cond*	T	totP*
2500-70 000	1.6-8.2	0.02-2.9	4.3-26.4

The relationship between model and empirical data is given in Fig. 3.25A. This is the best model obtained for logarithmic data on the effect parameter. But this is not to say that it must also be the best model for Cs-pi87 when given **in absolute values.** This is shown in Fig. 3.25B, where empirical data on Cs-pi87 are compared with model-calculated data on Cs-pi87 (from Eq. 3.7). The r^2-value has decreased from 0.91 to to 0.82. This depends on the fact that the regression in one case is done on absolute values whereas in the other it is done on logarithmic data. This influences the r^2-value. The best adaptation that can be made to non-logarithmic data on Cs-pi87 is given in Fig. 3.25C. Here, two new sensitivity parameters, CaMg and mire percent (Mire%), have been included. It is interesting to note that Mire% enters in this context. The larger the Mire% the greater the inflow of humus (i.e., the carrier particles for caesium), the greater the secondary dose and the higher the value of Cs-pi87. The hardness of the lake water (CaMg) naturally can be included in the same way as discussed earlier for cond and K.

The formula (Model 10) can be written:

$$\text{Cs-pi87} = -2769 + 0.07\text{Cs-soil} + 778.88T - 3004\log(\text{CaMg}^*) + 45.86\text{Mire}\% \qquad (3.8)$$
$$(r^2 = 0.85, \ n = 41)$$

where Cs-pi87=Cs-concentration in pike 1987 in Bq/kg ww; Cs-soil= fallout determined from Fig. 2.1, in Bq/m^2; CaMg*=mean hardness of lake water during the period March 1986 to February 1987 in meq/l; Mire%= percentage of mires in the catchment area; and T=theoretical water turnover time (in years).

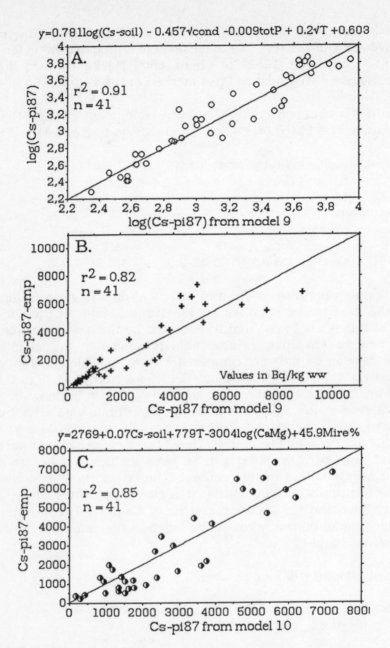

Fig. 3.25. Relationship between empirical values of caesium concentration in pike in 1987 and model values.

A. Logarithmic data, according to model 9, which gave the highest degree of explanation ($r^2=0.91$). The equation is based on Cs-soil, cond, totP and T. The figure gives the regression line and its equation.

B. Relationship between empirical data on Cs-pi87 and model data (model 9) for absolute values (not logarithmic).

C. Relationship between Cs-pi87-emp and model data from model 10.

Model conditions:

Cs-soil	T	CaMg*	Mire%
2500-70 000	0.02-2.9	0.1-0.64	0-35

It is important to note that the primary dose (Cs-soil) and the theoretical water turnover time (T) are included in these two models (3.7 and 3.8).

Since the three sensitivity parameters (T, CaMg and Mire%) in Eq. (3.8) emanate from different clusters/functional groups, it is also interesting to see what would happen to the predictability/r^2-value of the model if these parameters are **replaced** by other parameters from the same cluster. This is illustrated in Table 3.15. From this table, we may note:

Table 3.15. Which parameters could replace the three sensitivity parameters in Eq. (3.8)?

Mire%		T			CaMg*
Im%	0,19	**Vol**	**0,68**	cond*	0,97
Lake%	0,15	**area**	**0,59**	Ca*	0,95
Q	0,14	**Atot**	**0,59**	alk*	0,92
ADr	0,10	**Dm**	**0,56**	pH*	0,65
Mor%	-0,06	**Dmax**	**0,53**	totP*	0,61
Acid%	-0,13	Bm	0,50	Fe*	0,08
Br	-0,16	Lmax	0,48	Lake%	0,08
Ol%	-0,30	lo	0,44	Acid%	0,04
Rock%	-0,30	BA	0,30	Col*	-0,08
For%	**-0,86**	F	-0,05		
		DR	-0,19		
		Vd	-0,22		
		Q	-0,22		
		BET	-0,30		

	Parameter				r^2
Step 1	Cs-soil				0,61
Step 2	Cs-soil	area			0,75
Step 3	Cs-soil	area	log(totP)		0,79
Step 4	Cs-soil	area	log(totP)	For%	0,79

Cs-pi87 = 5437 + 0.073Cs-soil - 26.23For% - 3396log(totP) + 819.6area
       (r^2=0.79, n=41)

• Mire% can be replaced by For% (r=-0.86 between these two catchment area parameters).

• The volume (Vol), water area (a), total lake area (Atot), mean depth (Dm) and max. depth (Dmax), can, with decreasing relevance, replace theoretical water turnover time (T).

• Conductivity, Ca-concentration, alkalinity, pH, totP-concentration and Fe-concentration can, with decreasing relevance, replace hardness (CaMg).

If we, e.g., replace Mire% with the For%, water turnover time with lake area and hardness with totP, we get a formula that gives a 79% degree of explanation (see Table 3.15), which is only 6% lower than that given by Eq. (3.8).

> In this context, the main rule is to try to obtain relationships that are as simple as possible. If the main interest is to be able to predict values of Cs-pi87, then the regression analysis should be based on Cs-pi and not on log(Cs-pi).

3.2.1. MODEL STABILITY

It is always important to test the validity and stability of models, ecometric as well as dynamic. Here, we will give examples of what happens to model 9 (Eq. 3.7) if it is **tested in three different ways**:

• Firstly, with a "weight window" in the data material where only fish with a weight around 1 kg have been accepted (from 0.6-1.4 kg; the number of lakes then remaining is 32; here we only accept lakes where at least four pike have been analysed). The intention of the test is to see whether the r^2-value is then improved. The results are given in Fig. 3.26A. The r^2-value then decreases slightly from 0.91 to 0.87.

• In the next test, we have placed a "time window" in the data material and only accepted data from the spring months of April and May (we accept only lakes where four pike have been analysed; the number of lakes will then be 37). The intention of the test is to see whether the r^2-value can be improved if we standardize for time variations which may be present in the original material where the time of catching varies from January to June. The results are given in Fig. 3.26B. This figure shows that the regression line and r^2-value remain unaltered in comparison with those in Fig. 3.22.

• Finally, we have compared the results from Model 9 for 1987 with the caesium data for 1988. The results are shown in Fig. 3.26C (n=40). The results in this figure are interesting in several ways. First, we can note that the r^2-value is very high (0.89). The model for the 1987 material thus also

73

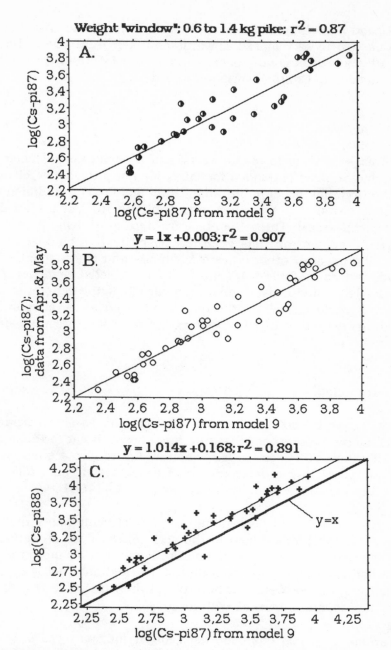

Fig. 3.26. How well can model 9 be used to predict the caesium concentration in pike?
A. Here a comparison is made of empirical data [log(Cs-pi87)] on the y-axis with model data on the x-axis when only pike in the weight interval 0.6-1.4 kg are accepted. The degree of explanation is 87%.
B. Here empirical data [log(Cs-pi87)] are compared with model data when only data from April and May are accepted. The degree of explanation is 91%.
C. Here empirical data from 1988 [log(Cs-pi87)] are compared with data from model 9 (i.e., data from 1987). The degree of explanation is 89%. y=x is shown by the solid line.

gives a very good description of the conditions during 1988. But the regression line has, so to say, moved upwards one step from 1987; the concentrations are thus throughout higher in pike in 1988 than they were in 1987. This brings us to Chap. 4, variation in time.

3.2.2. OTHER LOAD MODELS

The aim of this section is to give a very brief presentation of ecometric load models/diagrammes developed for lakes for non-radioactive elements (example mercury), for other pollutants than metals and radioisotopes (example phosphorus) and for other environments than lakes (example nitrogen in coastal areas). Thus, the objective here is to demonstrate that the same general set-up that has been discussed in this chapter for Cs-pi87 may also be used for other types of elements and environments. This set-up is summarized in Table 3.16. So, for these selected elements (Hg, P and N), and for lakes and coastal areas, we will follow the general ecometric scheme and try to answer to the questions given as seven points in Table 3.16 (from definition of effect parameter to model presuppositions and limitations).

3.2.2.1. Mercury in lakes

There are more than 50 000 publications on "Mercury as en environmental pollutant". Mercury does not constitute any known threat to the aquatic ecosystem. The aim of this section is, evidently, not to give a summary on all aspects connected to the development of an ecometric load model for mercury in lakes (see Håkanson, 1980; Lindqvist et al., 1991).

Figure 3.27 gives a 3-D load diagramme, the model behind this diagramme, and the presuppositions linked to the given effect, dose an sensitivity parameters. The environmental goal for Sweden, that the mercury concentration in fish from all of Sweden´s 83 000 lakes should not exceed the level of 0.5 mg Hg/kg wet weight, where the authorities place restrictions for human consumption, is marked with an arrow.

The effect parameter, which has been used for Hg for many years, is the Hg-concentration in 1 kg pike. The value depends on how the pike has lived and what it has eaten during a long period before being caught (the half-life for Hg in 1 kg pike is about 3 years). The Hg-concentration in pike is an integrated value for the entire environment of the pike and its prey. The Hg-content in pike depends on the Hg-dose supplied to the entire environment and not only to the clump of reeds where the pike was caught, as well as to the biological, chemical and physical conditions in this environment over a long period, since these conditions influence

Table 3.16. Summary of set-up to establish an ecometric load model and a load diagramme.

1. Define an **effect parameter**, and its area and time resolution. In this example, Cs-pi87 is used as an effect parameter, pike as indicator organism.
Area constant: entire lake (area < 10 km^2).
Time constant: about 2 years.

2. Define area and time **compatible dose and sensitivity parameters.**

3. Set up an ecometric matrix.

4. **Establish clusters and functional groups**, e.g. , by means of correlation analysis using linear correlation coefficients (r).

Example: r-rank table - Cs-pi87 versus different dose and sensitivity parameters.

Cs-pi87 Cs-param.		Cs-pi87 Water chem. param.		Cs-pi87 Morpho-metric param.		Cs-pi87 Catch-ment param.	
Cs-pi87	1	pH	-0.54	Q	0.58	Im%	0.69
Cs-wa	0.87	CaMg	-0.36	Dm	0.38	NBr	0.6
Cs-su87	0.67	alk	-0.36	Vol	0.08	ADr	0.57
Cs-su86	0.79	Ca	-0.35	T	0	Rock%	0.49
Cs-soil	0.76	totP	-0.35	Area	-0.14	Bas%	0.28
Cs-bo87	0.77	cond	-0.3	BA	-0.49	Mire%	0.07
Cs-bo86	0.63	K	-0.3	DR	-0.63	Forest%	0.07
		Fe	-0.2			Lake%	0.02
		Colour	-0.08			Br	0
						Coar%	-0.04
						Fine%	-0.07
						Mor%	-0.19
						Ol%	-0.2
						Acid%	-0.7

5. Establish an ecometric model, e.g. ,by means of step-wise multiple regression analysis; compare results against empirical data; give r^2-value and number of data-pairs.

	Parameters	r^2
Step 1	Cs-soil	0.59
Step 2	Cs-soil, cond	0.77
Step 3	Cs-soil, cond, T	0.86

Model (in this example): Cs-pi87=1233+0.073Cs-soil-510cond+893T; (r^2=0.86; n=41)

6. Construct a load diagram; note critical loads, guidelines, etc.

7. Define **presuppositions and limitations** ("traffic rules").

Cs-soil (Bq/m^2)	cond (mS/m)	T (yr)
2500-70 000	1.6-8.2	0.02-2.9

This model is **not** applicable for other species of fish, other time periods and other lake types.

Load diagramme - Mercury in lakes

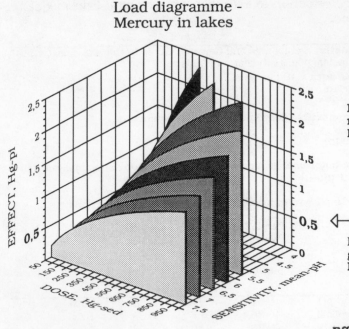

Diagram valid for mesotrophic lakes - BPI = 3

⟵ _____

National environmental goal in Sweden:
Hg-pi < 0.5

Model (Håkanson, 1980):

Hg-pi=(4.8*log(1+Hg50/200))/((pH-2)*log(BPI))
(r^2=0.74; n=18)

Legend:

Dose parameter: Hg50=median Hg-content in mg Hg/kg dw
from surficial (0-1 cm) sediment samples
Sensitivity parameters: 1. pH=yearly mean lake pH
2. BP=bioproduction index; based on surficial sediment data on
      nitrogen content and organic content (=loss on ignition)

Presuppositions:

80<Hg50<2040
4.8<pH<7.6
2.9<BPI<5.2

Model not applicable:

• For lakes smaller than 1 km^2

• For dystrophic lakes with IG larger than 25% (IG=organic
  content=loss on ignition in % dw)

Effect parameter for mercury in lakes

Hg-pi: Hg-content in muscle
in 1 kg pike in mg/Hg/kg ww
(at least 5 fish in the size
range from 0.5 to 1.5 kg from
each lake should be analysed)

• area constant: entire lake for
lakes smaller than 25 km^2

• time constant: ≈ 3 years

All dose and sensitivity
parameters should have the
same area and time resolution
as the effect parameter

Other effect parameters:

No (except Hg-content in
muscle from other species of
fish for human consumption)

Fig. 3.27. Load diagramme for mercury in lakes; model; presuppositions.

the distribution of the Hg-dose on different types of "carrier particles"
(like humus, clays, etc) and the bioavailability of the Hg-dose. The pH of
the water is important for the binding of mercury to different types of
carrier particles and for how Hg is distributed among different forms,
such as Hg^0, Hg^+ and methyl-Hg.

In the model, pH explains the greatest proportion of the variation in the effect parameter (Fig. 3.28). The next most important factor, according to the stepwise regression, is the Hg-dose to the lake (Hg50, i.e., the median Hg-content of surficial sediment samples). It is interesting to note that a sensitivity factor can, in fact, be more important than a dose factor in explaining the variation in an effect parameter in nature. If pH and Hg50 are pooled, 63% of the variation in Hg-pi among lakes can be statistically explained. If one further sensitivity factor is added, the degree of bioproduction (BPI), then the degree of explanation increases to 74%. This is as far as one can proceed with this particular set of data (see Håkanson, 1980). The latter link is also interesting since it demonstrates that the Hg-concentration in fish is not only increased by low pH/acidification (**synergism**, i.e., pollution of acidifying substances like S increases the effect, here given by values on Hg-pi), but also linked to another major environmental problem in aquatic systems, namely eutrophication. The greater the BPI the greater the production of algae, plankton and fish, and the greater the amount of biomass in the lake. This implies that a given Hg-dose is distribution over a larger biomass whereupon the Hg-content in the biomass becomes less (**antagonism**, i.e., pollution of nutrients like P and N decrease the effect parameter). This is one explanation why the BPI value is such an important factor in this context. Another explanation concerns the internal correlations: If the bioproduction in a lake is increased, then the entire character of the ecosystem is changed, which influences practically all other water chemical and biological factors with which a ecosystem can be described.

Figure 3.28 illustrates how this particular ecometric model may be used in remedial contexts in a highly polluted (Hg50=680 mg Hg/kg; a natural background level for Hg in sediments would be about 100-150) and severely acidified (pH=4.8) lake. On the basis of this model, one may determine which measures could be used in order to decrease the Hg-content in fish:

- The direct emissions of Hg to the lake could be halted (step 2 in the figure).
- The atmospheric emissions could be reduced; this would cause a reduction in the atmospheric deposition and, in due course, also in the secondary load from the catchment (step 4 in the figure; note the very long lag-phase between emission reduction and reduction of Hg in fish in this case).
- The pH could be increased (e.g., by liming; steps 1, 3 and 5 in the figure), and
- The bioproduction could be increased (e.g., by fertilization, i.e., the adding of nutrients, predominantly phosphorus).

Naturally, the bioproduction must not be increased indiscriminately in order to reduce the Hg-content in fish. This could lead to a eutrophication problem! But this is not any particular problem since high Hg-contents in fish mainly occur in low-productive (oligotrophic) lakes. In such lakes, it may be practical, economical and ecologically relevant to lime the lakes in order to increase the pH (see Håkanson et al., 1990b). Naturally, the best way in the long run to reduce Hg-concentrations in fish is to stop emissions of Hg.

EXAMPLE – PRACTICAL USE OF LOAD MODELS FOR Hg IN LAKES

A. Model Håkanson (1980): Only applicable under specified conditions ("traffic rules").

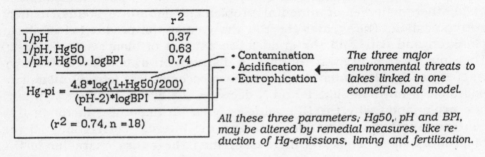

	r^2
1/pH	0.37
1/pH, Hg50	0.63
1/pH, Hg50, logBPI	0.74

$$Hg\text{-pi} = \frac{4.8*\log(1+Hg50/200)}{(pH-2)*\log BPI}$$

$(r^2 = 0.74, n = 18)$

- Contamination
- Acidification
- Eutrophication

The three major environmental threats to lakes linked in one ecometric load model.

All these three parameters; Hg50, pH and BPI, may be altered by remedial measures, like reduction of Hg-emissions, liming and fertilization.

B. **LOAD DIAGRAM**

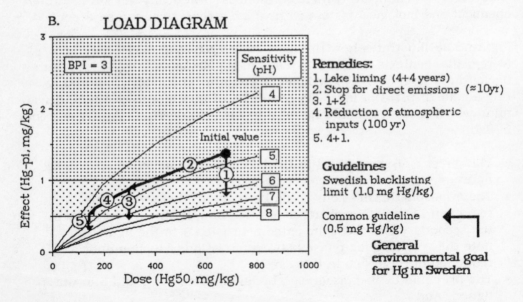

Remedies:
1. Lake liming (4+4 years)
2. Stop for direct emissions ($\approx$10yr)
3. 1+2
4. Reduction of atmospheric inputs (100 yr)
5. 4+1.

Guidelines
Swedish blacklisting limit (1.0 mg Hg/kg)

Common guideline (0.5 mg Hg/kg)

General environmental goal for Hg in Sweden

Fig. 3.28. Practical use of load models and diagrammes in contexts of remediation.

3.2.2.2. Phosphorus in lakes

Numerous publications exist on lake eutrophication. The famous Vollen-weider model (Vollenweider, 1968, 1976; and later versions, e.g., OECD, 1982), and the analysis behind this load model, constitutes a fundamental base for practically all environmental assessments of phosphorus in limnic environments. In models of this type, the effect parameter is generally linked to the concepts of hypertrophy (extremely productive lakes), eu-trophy, mesotrophy and oligotrophy, which are related to defined phos-phorus concentrations in the lake water. Many other effect parameters are discussed in contexts of lake eutrophication, e.g., focussed on the fish and bottom fauna community, algae, primary productivity, Secchi depth or the oxygen conditions (see Fig. 3.29). The linkage between nutrient dose, lake sensitivity and environmental effect parameter is illustrated as a 3-D load diagramme in Fig. 3.29. The model behind this diagramme, and the presuppositions linked to the given effect, dose and sensitivity parameters are given in the same standardized way as previously for Cs-137 and mercury.

3.2.2.3. Nitrogen in coastal areas

It is not possible to make simple adjustments of, e.g., the Vollenweider model to marine areas, because for such areas nitrogen is often the most limiting nutrient for bioproduction (see Rosenberg, 1986; Ambio, 1990) and because the physical conditions in the sea often have a profound im-pact on the biological conditions in a given coastal area.

Figure 3.30 gives a load diagramme/model for nitrogen in marine coastal areas. This particular model has been developed for evaluations of environ-mental effects in coastal areas of the Baltic affected especially by emissions from point sources (like fish farms).

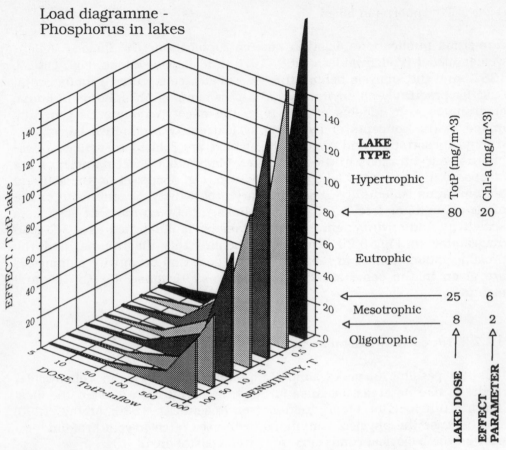

Load diagramme -
Phosphorus in lakes

Model, OECD (1982):
TotP-lake=1.55TotP-in/(1+√T)^0.82)

Legend and presuppositions:
<u>Dose parameter</u>: TotP-inflow (in mg/m^3); range: 1-150

<u>Sensitivity parameter</u>: Lake water retention time, T
(in years); range: 0.1-100

<u>Effect parameter</u>: TotP-lake (in mg/m^3); range: 2.5-100

Model:
Degree of explanation: r^2=0.86
Number of lakes: 87

Other effect parameters:

• HOD (hypolimnetic oxygen demand)
• Secchi depth
• Algal volume
• Fish community index
• Bottom fauna index

Model not applicable for:
• Monomictic lakes (model only applicable for
  lakes from temperate climates)
• Dams/reservoirs
• Lakes with high internal loading

Fig. 3.29. Load diagramme for phosphorus in lakes; model; presuppositions; linkages between traphic status (oligotrophy to hypertrophy), totP-concentrations and concentrations of chlorophyll-a in lake water; other effect parameters concerning lake eutrophication.

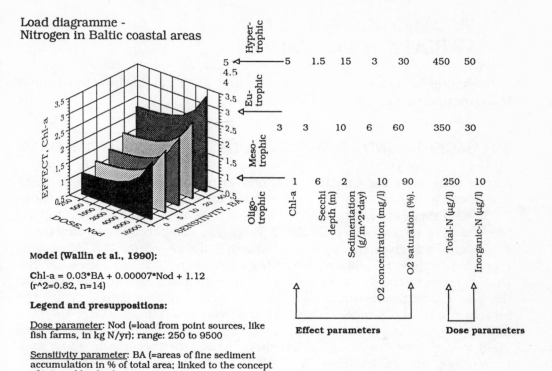

**Load diagramme -
Nitrogen in Baltic coastal areas**

Model (Wallin et al., 1990):

Chl-a = 0.03*BA + 0.00007*Nod + 1.12
(r^2=0.82, n=14)

Legend and presuppositions:

Dose parameter: Nod (=load from point sources, like
fish farms, in kg N/yr); range: 250 to 9500

Sensitivity parameter: BA (=areas of fine sediment
accumulation in % of total area; linked to the concept
of internal loading); range: 0 to 42

Effect parameter: Mean chlorophyll concentration in
mg/m^3, Chl-a:
• area constant: 1 to 14 km^2
• time constant: 2 to 6 weeks during the production
period (May-Sept.)

All dose and sensitivity parameters should have the
same area and time constants, i.e., they should be mean
values from coastal areas of this size and for similar
time periods.

Model not applicable for:

•Lakes, brackish waters and estuaries (where also
the load of phosphorus has to be accounted for)
• Marine areas influenced by tides (dH>25 cm)
• Coastal areas outside the range of the model

*Fig. 3.30. Load diagramme for nitrogen in Baltic coastal areas; model;
presuppositions; linkages to other effect parameters of coastal eutrophication and to two (indirect) dose parameters (concentrations of
total-N and inorganic-N in Baltic coastal areas).*

4. ECOMETRIC MODELLING — VARIATION IN TIME

This chapter is meant to give examples and exercises on how to use and abuse ecometric models.

4.1. BACKGROUND TO THE CALCULATIONS - EMPIRICAL DATA

Figure 4.1 shows the Cs-concentrations in pike in our 41 lakes in 1987 and 1988. It should be noted that all data for 1988 are higher than those for 1987. From Fig. 4.1, it is difficult to get a clear picture of what is going on. The lake mean values for the 2 years are compared more directly in Fig. 4.2, which gives the regression line and the r^2-value. The lakes with high Cs-pi values in 1987 also had high values in 1988. In Fig. 4.3A, the change between the years, as a function of the values for 1987, is given on the y-axis. The figure shows that there is a clear relationship between the increase (dCs-pi=Cs-pi88 - Cs-pi87) and the values from 1987, i.e., the higher the Cs-concentration in pike in 1987, the larger the increase to 1988 in absolute values. If, on the other hand, the increase is given as a percentage, we can see from Fig. 4.3B that the increase is independent of the starting value (i.e., Cs-pi87). The average percentage increase between the 2 years is, according to Fig. 4.3B, as much as 70%.

Figure 4.4A shows the average development in 1+perch during the 3 years. **The lake mean values decrease from year to year as well as the variation between the lakes**. Small perch attains a different niche in the lake eco-system than pike; it is a small plankton-eating fish and is a prey hunted by pike.

Figure 4.4B shows how the Cs-concentrations in material from sediment traps placed close to the bottom have decreased during the 3 years. Figure 4.4C illustrates how the caesium deposition (the flow in $Bq/m^2/90$ days during the production period) has varied in bottom traps. This figure is based on data on the length of the registration period, the deposition of material and the Cs-concentration, and the resulting figures have been standardized to 90 days. As can be seen from the diagramme, the Cs-flow also shows a marked decrease during the 3 years; the variation between the lakes also decreases.

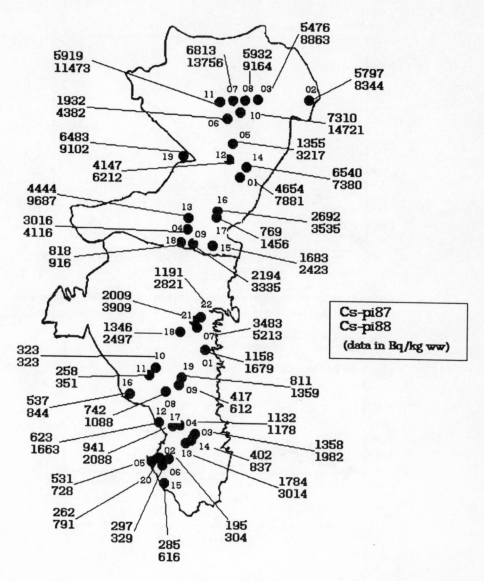

Fig. 4.1. Caesium concentrations in pike in 1987 (upper values) and 1988 (lower values) for our 41 lakes in the provinces of Gävleborg and Västernorrland. The province code of the lakes is given.

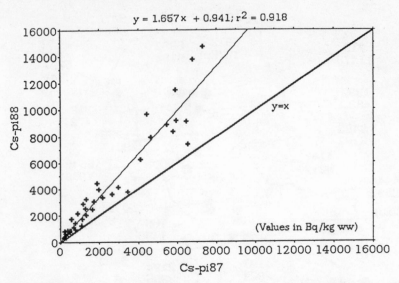

Fig. 4.2. Comparison between caesium concentrations in pike in 1987 and in 1988 from our 41 lakes. The regression line, the r^2-value and the y=x line are given.

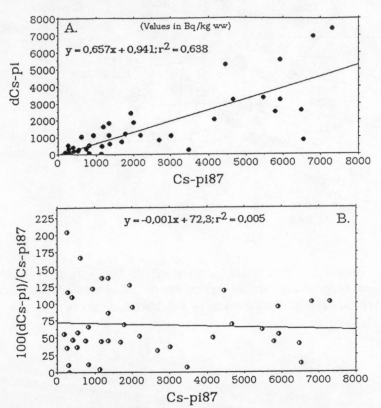

Fig. 4.3. A. Change in Cs-pi between 1987 and 1988 (y-axis) as a function of the values in 1987 (x-axis). The regression line and the r^2-value are given.
B. The percentage change in Cs-pi between 1987 and 1988 (y-axis) as a function of the values in 1987 (x-axis). The regression line and the r^2-value are given.

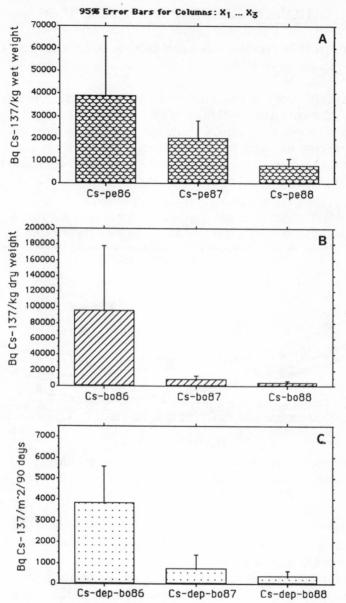

Fig. 4.4. A. The change (decrease) in Cs-concentration between 1986, 1987 and 1988 in:

A. 1+perch

B. Material in sediment traps placed close to the bottom.

C. Caesium deposition in sediment traps placed close to the bottom.

4.2. CALCULATION OF CAESIUM IN WATER 1986

☞ **The task in this section has been to calculate what the Cs-concentrations were in water in 1986.**

Figure 4.5 shows how the Cs-concentrations vary with time after the Chernobyl event (month 1 then refers to May 1986, etc.) in the 15 best studied lakes where we have data on Cs from sediment traps and lake water. The figure shows that we only have empirical data from month 13 (i.e., May 1987) and that, subsequently, the Cs-concentrations decrease or are fairly constant. This is in fair agreement with Fig. 4.4.

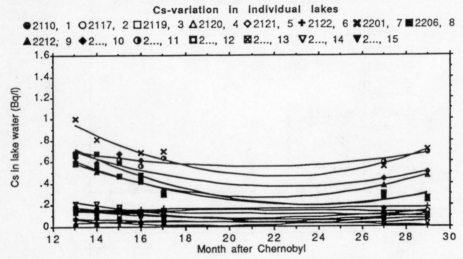

Fig. 4.5. The variation in caesium concentration in lake water in our 15 lakes from month 13 (May 1987) to month 29 after the Chernobyl event.

The calculations here have been made in three steps. The aim has been to consider differences between the catchment areas of the lakes and the lake morphometry and in that way attempt to obtain a picture of the environmental characteristics that are of particular importance in controlling the recovery process as regards the caesium dose to the lakes.

Step 1
The first step has been to try to link the values for caesium in water in 1987 and 1988 to data on caesium in bottom sediment traps. A correlation matrix is given in Table 4.1. Table 4.1 shows that there is a strong link between caesium in water and caesium in bottom sediment traps (r=0.77). There are also marked relationships between caesium in water and several environmental parameters, mainly Acid%, i.e., the percentage of acidic bedrocks, the area of accumulation bottoms (BA), the mean depth (Dm) and the water discharge (Q).

Table 4.1. Correlation matrix for Cs-wa87/88 vs lake morphometric parameters and parameters describing the catchment area.

	Cs-wa87/88	Cs-bo87/88	area	Vol	Dm	DR	BA	Q	T	ADr	Br	Lake%	For%	Mire%	Ol%	Rock%	Mor%	Im%	Acid%
Cs-wa87/88	1,00	0,77	0,00	0,26	0,58	-0,67	-0,62	0,52	0,04	0,51	-0,13	0,12	0,07	0,05	-0,20	0,47	-0,15	0,75	-0,75
Cs-bo87/88	0,77	1,00	-0,10	0,05	0,35	-0,49	-0,38	0,18	0,12	0,18	-0,11	0,24	0,06	-0,25	0,17	0,51	-0,35	0,63	-0,63
area	0,00	-0,10	1,00	0,93	0,52	0,16	0,23	0,17	0,34	0,14	-0,32	-0,01	-0,40	0,64	-0,29	-0,51	0,32	0,02	-0,01
Vol	0,26	0,05	0,93	1,00	0,77	-0,16	-0,07	0,29	0,30	0,27	-0,30	-0,04	-0,29	0,55	-0,32	-0,34	0,28	0,19	-0,19
Dm	0,58	0,35	0,52	0,77	1,00	-0,62	-0,61	0,33	0,12	0,33	-0,25	0,06	-0,08	0,22	-0,21	0,14	0,01	0,45	-0,46
DR	-0,67	-0,49	0,16	-0,16	-0,62	1,00	0,60	-0,28	0,00	-0,28	-0,02	-0,19	-0,02	0,11	-0,03	-0,63	0,36	-0,64	0,64
BA	-0,62	-0,38	0,23	-0,07	-0,61	0,60	1,00	-0,25	0,10	-0,26	0,09	0,01	-0,34	0,24	0,15	-0,31	-0,02	-0,32	0,34
Q	0,52	0,18	0,17	0,29	0,33	-0,28	-0,25	1,00	-0,49	-0,43	0,01	0,01	0,33	-0,20	0,14	-0,09	0,39	-0,39	-0,39
T	0,04	0,12	0,34	0,30	0,12	0,00	0,10	-0,49	1,00	-0,51	0,15	-0,13	0,21	-0,04	-0,21	-0,42	0,44	-0,04	0,05
ADr	0,51	0,18	0,14	0,27	0,33	-0,28	-0,26	-0,43	-0,51	1,00	-0,43	0,02	-0,18	0,30	-0,18	0,16	-0,11	0,39	-0,39
Br	-0,13	-0,11	-0,32	-0,30	-0,25	-0,02	0,09	0,01	0,15	-0,43	1,00	-0,36	-0,09	0,00	0,28	0,10	-0,16	-0,23	0,24
Lake%	0,12	0,24	-0,01	-0,04	0,06	-0,19	0,01	0,01	-0,13	0,02	-0,36	1,00	-0,44	-0,04	0,32	0,47	-0,39	0,47	-0,37
For%	0,07	0,06	-0,40	-0,29	-0,08	-0,02	-0,34	0,33	0,21	-0,18	-0,09	-0,44	1,00	-0,66	-0,39	-0,15	0,51	-0,24	0,22
Mire%	0,05	-0,25	0,64	0,55	0,22	0,11	0,24	-0,20	-0,04	0,30	0,00	-0,04	-0,66	1,00	-0,40	-0,23	0,06	0,09	-0,09
Ol%	-0,20	0,17	-0,29	-0,32	-0,21	-0,03	0,15	0,14	-0,21	-0,18	0,28	0,32	-0,39	-0,40	1,00	0,37	-0,66	0,11	-0,09
Rock%	0,47	0,51	-0,51	-0,34	0,14	-0,63	-0,31	-0,09	-0,42	0,16	0,10	0,47	-0,15	-0,23	0,37	1,00	-0,79	0,65	-0,65
Mor%	-0,15	-0,35	0,32	0,28	0,01	0,36	-0,02	0,39	0,44	-0,11	-0,16	-0,39	0,51	0,06	-0,66	-0,79	1,00	-0,49	0,47
Im%	0,75	0,63	0,02	0,19	0,45	-0,64	-0,32	-0,39	-0,04	0,39	-0,23	0,47	-0,24	0,09	0,11	0,65	-0,49	1,00	-1,00
Acid%	-0,75	-0,63	-0,01	-0,19	-0,46	0,64	0,34	-0,39	0,05	-0,39	0,24	-0,37	0,22	-0,09	-0,09	-0,65	0,47	-1,00	1,00

Step 2

Table 4.2 gives the result of the step-wise multiple regression analysis. The analysis has been made on the material from 1987 and 1988, i.e., on 15+15=30 lakes (see Table 4.3). The "constant" environmental parameters do not change between years but the Cs-bo values do change. The most important individual parameter in explaining variations between lake data on Cs-wa87/88 is Cs-bo87/88; the r^2-value is 0.59.

Table 4.2. Results from step-wise multiple regression analysis to statistically explain the variation in Cs-wa87/88 between lakes: The regression model that yields the highest degree of explanation for one dose (Cs-bo87/88) and three sensitivity parameters (Q, Dm and Br) is given as Eq. (4.1).

	Parameter	r^2-value
Step 1	Cs-bo87/88	0.59
Step 2	Cs-bo87/88, Q^2	0.75
Step 3	Cs-bo87/88, Q^2, Dm^2	0.85
Step 3	Cs-bo87/88, Q^2, Dm^2, log(Br)	0.87

Logically, the next most important parameter is water discharge (Q). If this parameter is also included, the r^2-value increases from 0.59 to 0.75. The third most important parameter is mean depth (Dm). This parameter is directly linked to the bottom dynamic conditions. The fourth factor is the relief of the catchment area (Br). The model that is based on these four parameters gives a 87% degree of explanation for the variation in mean Cs-concentration in lake water in 1987 and 1988. The regression line is given by:

$$Cs\text{-}wa87/88 = 0.0000217 Cs\text{-}bo87/88 + 2.53 Q^2 + 0.005 Dm^2 + 0.188 \log(Br) - 0.363 \quad (4.1)$$
$$(r^2 = 0.87, \; n = 30)$$

where Cs-wa87/88=Cs-concentration in lake water in 1987 and 1988; mean value for the period May-Sept. (Bq/l); Cs-bo87/88=Cs-concentration in bottom sediment traps during May-Sept. (Bq/kg dry matter); Dm=mean depth (m); Br=relief of the catchment area.

The relationship applies for lakes when Cs-bo, Q, Dm and Br fall within the following ranges:

Cs-bo	Q	Dm	Br
310-23 100	0.01-0.26	2.1-8.3	34-194

The relationship between empirical data on Cs-wa and model data from Eq. (4.1) is given in Fig. 4.6. Equation (4.1) has only been used to calculate backwards, i.e., to obtain data on Cs-wa86. The equation should not be used to predict the future development of Cs-wa since this would, e.g., require data on Cs-bo **for future years** (these must then first be predicted). In addition, with this equation, Cs-wa would not become zero if Cs-bo was zero, which it should be. In a following section, we will discuss models for prediction of Cs-wa in future years.

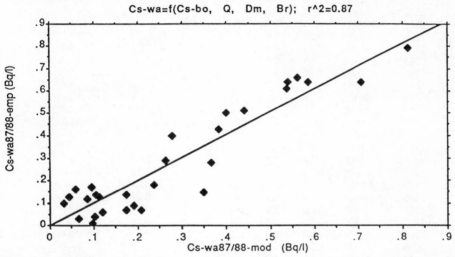

Cs-wa=f(Cs-bo, Q, Dm, Br); r^2=0.87

Fig. 4.6. Comparison between empirical data (y-axis) on caesium in lake water in 1987 and 1988 and model-calculated data (according to Eq. 4.1) on the x-axis. The model gives a r2-value of 0.87 (n=30).

Step 3
On the basis of the relationship shown in Eq. (4.1), by entering empirical data for **Cs-bo86**, we can calculate what the Cs-concentrations in lake water should have been in our 15 lakes in 1986. The results are given in Table 4.3 and are summarized in Fig. 4.7. It is clear that also the Cs-concentrations in water decreased markedly from 1986 to 1987 and continue to decrease, but more slowly. The highest value calculated in this way for 1986 is about 13 Bq/l (lake 2201), which is a very high figure. This depends on the extreme value for this lake as regards Cs-bo86 of 576 000 Bq/kg dry weight.

In summary, we may state that the Cs-concentrations in lake water were very high in 1986 in these lakes, that there is a clear link between Cs-wa and Cs-bo and that this link mainly depends on the tributary water discharge (Q), the mean depth of the lakes (Dm) and the relief of the catchment area (Br). These relationships appear logical: Q and Br concern the water transport from land to water, and Dm concerns the

Table 4.3. Calculation of caesium concentrations in water in 1986 and a comparison between model-calculated values and empirical values.

Lake	Cs-bo (Bq/kg dw)	Q (m3/s)	Dm (m)	Br	Cs-wa-emp (Bq/l)	Cs-wa-mod (Bq/l)
1986						
2110	8400	0,13	4,7	34	•	0,26
2117	36700	0,15	4,2	34	•	0,87
2119	19200	0,01	2,7	60	•	0,42
2120	3300	0,03	2,1	44	•	0,04
2121	37700	0,02	2,4	194	•	0,91
2122	29800	0,04	2,8	84	•	0,69
2201	576000	0,20	8,3	56	•	12,90
2206	68100	0,26	6,4	46	•	1,80
2212	241000	0,18	4,9	53	•	5,39
2213	137000	0,05	5,1	80	•	3,10
2214	115000	0,39	2,9	42	•	2,86
2215	95100	0,08	4,5	80	•	2,17
2216	56200	0,06	4,0	45	•	1,26
2217	19100	0,11	3,8	62	•	0,49
2218	6400	0,17	2,1	54	•	0,20
1987						
2110	1060	0,13	4,7	34	0,04	0,10
2117	670	0,15	4,2	34	0,12	0,09
2119	1600	0,01	2,7	60	0,13	0,04
2120	4400	0,03	2,1	44	0,03	0,07
2121	410	0,02	2,4	194	0,14	0,11
2122	800	0,04	2,8	84	0,16	0,06
2201	18520	0,20	8,3	56	0,79	0,81
2206	5312	0,26	6,4	46	0,51	0,44
2212	10890	0,18	4,9	53	0,50	0,40
2213	20973	0,05	5,1	80	0,64	0,59
2214	15590	0,39	2,9	42	0,64	0,71
2215	10878	0,08	4,5	80	0,15	0,35
2216	23145	0,06	4,0	45	0,61	0,54
2217	7330	0,11	3,8	62	0,18	0,24
2218	5246	0,17	2,1	54	0,07	0,17
1988						
2110	930	0,13	4,7	34	0,01	0,10
2117	1921	0,15	4,2	34	0,13	0,11
2119	1102	0,01	2,7	60	0,10	0,03
2120	311	0,03	2,1	44	0,07	-0,02
2121	3450	0,02	2,4	194	0,14	0,17
2122	2486	0,04	2,8	84	0,17	0,10
2201	6907	0,20	8,3	56	0,66	0,56
2206	1859	0,26	6,4	46	0,28	0,37
2212	5271	0,18	4,9	53	0,40	0,28
2213	11534	0,05	5,1	80	0,43	0,38
2214	7890	0,39	2,9	42	0,64	0,54
2215	3582	0,08	4,5	80	0,09	0,19
2216	10450	0,06	4,0	45	0,29	0,26
2217	6012	0,11	3,8	62	0,07	0,21
2218	2803	0,17	2,1	54	0,06	0,12
Max.	576000	0,39	8,3	194		
Min.	311	0,01	2,1	34		

bottom dynamics/resuspension activity. All other factors tested appear to be of less importance. The factors that are intercorrelated to these factors may replace them in the model, e.g., the theoretical water turnover time (T) and the size of the catchment area (ADr) could replace Q and Br, and the dynamic ratio (DR) and the area of accumulation bottoms (BA) could replace Dm, but in such situations the adaptation will not be as good.

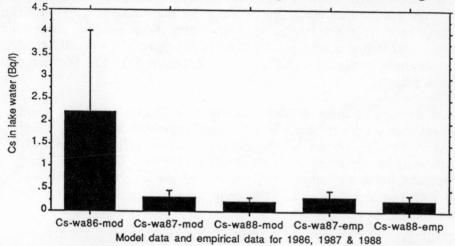

Fig. 4.7. Empirical data and model-calculated data on caesium in lake water (mean values and 95% confidence interval) for 1986, 1987 and 1988 in our 15 lakes.

4.3. RECOVERY

4.3.1. CAESIUM IN SEDIMENT TRAPS

☞ **The exercise here has been to predict the future development for caesium in material from sediment traps placed close to the bottom.**

Step 1
We will first test whether the decrease is best described by:

☞ (1) a simple **linear** function (**y=ax+b**).

☞ (2) a physical **exponential** function $(A(t)=A(O)*e^{-k*t}$, i.e. $t=(1/k)*\ln A(O) - (1/k)*\ln A(t))$, where the starting year, t0, is placed at **1986**; A(O)=Cs-activity year 0, t=time in years or months, and k=the time constant; or

☞ (3) the same physical function but where the starting year is placed at **1987** owing to the very special conditions prevailing during the Chernobyl year, 1986.

Which approach describes reality best? In some years, when more data are available, this question can be answered much better than today. The main object here is to discuss principles for how the decrease can be described using ecometric analysis.

Table 4.4 shows test results for caesium in bottom sediment traps for the three approaches. In this table, the r^2-value for each lake is given and the year when the extrapolated regression line cuts the x-axis, i.e., the year when the Cs-bo value is zero (**or 1**). The three approaches are illustrated for two randomly chosen lakes (2217 and 2218) in Fig. 4.8. We can see from Fig. 4.8 that:

Table 4.4. Testing of three different approaches (linear approach, exponential approach with startingyear 1986, and exponential approach with startingyear 1987) to describe the caesium concentrations in material from bottom sediment traps in 1986, 1987 and 1988. The intercepts of the regression lines, the slope coefficients and the r^2-values are given, together with the "zero" year, i.e., the year when the regression line cuts the x-axis.

| | Linear approach | | | | Exp. approach - start 1986 | | | | Exp. approach - start 1987 | | | |
Lake	r^2	Intercept	Slope	Zero yr from 1986	r^2	Intercept	Slope	Zero yr from 1986	r^2	Intercept	Slope	Zero yr from 1986
2110	0,76	7198	3735	1,93	0,79	8,71	1,10	8,71	-	6,97	0,13	53,18
2117	0,34	2962	875	3,39	0,14	7,75	0,32	23,92	-	6,51	-	-
2119	0,77	16350	9049	1,81	0,85	9,51	1,43	6,66	-	7,38	0,37	19,78
2120	0,50	4165	1495	2,79	0,66	8,59	1,18	7,27	-	8,39	2,65	3,17
2121	0,68	30978	17125	1,81	0,28	9,43	1,20	7,88	-	6,02	-	-
2122	0,70	24686	13657	1,81	0,45	9,51	1,24	7,66	-	6,69	-	-
2201	•	-	-	-	-	-	-	-	-	9,83	0,99	9,97
2206	0,79	58211	33121	1,76	0,95	10,88	1,80	6,04	-	8,58	1,05	8,17
2212	0,77	203585	117865	1,73	0,89	12,00	1,91	6,28	-	9,30	0,73	12,80
2213	0,81	119235	62733	1,90	0,92	11,62	1,24	9,39	-	9,95	0,60	16,64
2214	0,80	99715	53555	1,86	0,93	11,43	1,34	8,53	-	9,65	0,68	14,18
2215	0,81	82279	45759	1,80	0,97	11,29	1,64	6,88	-	9,29	1,11	8,37
2216	0,94	52807	22875	2,31	1,00	10,92	0,84	12,99	-	10,05	0,80	12,64
2217	0,83	17358	6544	2,65	0,87	9,73	0,58	16,84	-	8,90	0,20	44,95
2218	0,96	6615	1799	3,68	0,92	8,84	0,41	21,39	-	8,57	0,63	13,66
MV	0,75			2,19	0,77			10,42				18,13
SD	0,16			0,62	0,27			5,57				14.53

• With the first approach, we get an extrapolated value on the x-axis of 2-4 years for the two lakes, i.e., an **unrealistically rapid** recovery. This also applies for the other lakes, which can be seen from Table 4.4 where the time when the regression line cuts the x-axis is given for all 15 lakes. The mean value has been calculated to 2.2 years, i.e., by mid-1988 the concentrations of Cs-bo would have been approaching zero on average. Consequently, this approach cannot be used.

• With approach 2, a slower decrease is obtained. The extrapolated value for the two lakes in Fig. 4.8 is between 15 and 20 years. Figure 4.4 shows how this "zero year" varies between the lakes. The mean value is 10.4 years, i.e., in about 1997 the Cs-concentrations will, on average, be about zero in our 15 lakes. This is probably more realistic.

• With approach 3, an even slower recovery is obtained. In the two lakes the extrapolated values will be between 14 and 40 years. On average, it will take about 18 years to reach the zero level in our 15 lakes using this approach. However, since it is only based on two data inputs, the result is extremely unreliable.

Table 4.4 shows that approach 2 gives a slightly higher average r^2-value than approach 1 (0.77 and 0.75, respectively). No r^2-value can be determined for approach 3 since n=2.

It is probable that the recovery process can best be described using approach 2 but with a delay factor. The data on the "zero years" from approach 3 (Table 4.4) indicate that this delay factor should be 1.5-2. In subsequent sections, we will make calculations using approach 2 together with a delay factor of 1.5. This means that the "zero year" would happen 15 years after 1986 rather than 10 years, as indicated by approach 2, or 18 years, as indicated by approach 3.

The Cs-concentration in bottom sediment traps in a randomly chosen year after the Chernobyl event can then be calculated from the equation:

$$Cs\text{-}bo(t) = Cs\text{-}bo86 * e^{-kbo * t} \qquad (4.2)$$

where Cs-bo(t)=the Cs-concentration in bottom sediment traps in year t (Bq/kg dw); Cs-bo86=the Cs-concentration in bottom sediment traps in 1986 (Bq/kg dw); t=months after the Chernobyl event - August 1986=4, August 1987=16, etc; kbo=**the rate constant** for Cs-bo in each lake.

In order to determine the actual Cs-bo(t), e.g., for year 2 000 (Cs-bo00), we thus need data on kbo for the lake, as well as a starting value, Cs-bo86.

Step 2 - kbo
Here we have investigated which "constant" environmental factors influence kbo. Table 4.5A shows an r-ranking table of the factors that co-vary with kbo. From the table, we can see that the dynamic ratio (DR), the area of A-bottoms (BA), the mean depth (Dm), the percentage of open land in the

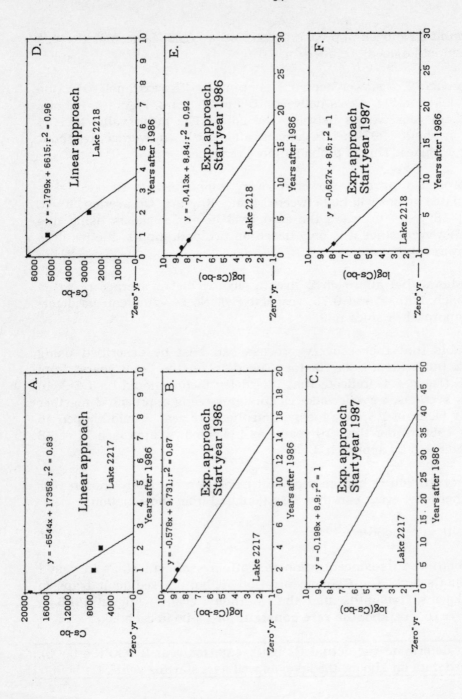

Fig. 4.8. Upper left: The linear approach for lake 2217.
Centre left: The exp. approach with startingyear 1986 for lake 2217.
for lake 2218.
Lower left: The exp. approach for startingyear 1987 for lake 2217.
The figures give the regression lines and the r^2-values.

Upper right: The linear approach for lake 2218.
Centre right: The exp. approach with startingyear 1986
Lower right: The exp. approach for startingyear 1987 for
lake 2218.

catchment area (Ol%) and the forest percent (For%) have high r-values with respect to kbo (r=-0.64, -0.59, -0.59, -0.49 and +0.46). A negative r-value in Table 4.5 (as for DR) implies that the Cs-concentrations in sediment traps decrease slowly with time; the results, in terms of the various steps, are summarized in Table 4.5B.

Table 4.5. A. Table showing r-ranking for kbo.

	kbo		kbo
Kbo	1.00	area	-0.05
Dm	0.59	Mire%	-0.06
For%	0.46	Acid%	-0.22
Vol	0.26	Lake%	-0.26
Q	0.20	Ol%	-0.49
ADr	0.20	BA	-0.59
Im%	0.19	DR	-0.64
Mor%	0.18		
Rock%	0.17		
T	0.09		
Br	0.08		

B. The "ladder".

	Parameters	r^2-value
Step 1	log(DR)	0.43
Step 2	log(DR), Ol%	0.68
Step 3	log(DR), Ol%, log(Br)	0.77
Step 4	log(DR), Ol%, log(Br), For%	0.80

The best adaptation with these four environmental parameters is:

$$kbo = -0.273 + 0.001For\% - 0.004Ol\% - 0.2\log(DR) + 0.069\log(Br) \quad (4.3)$$
$$(r^2=0.80, \ n=15)$$

where kbo=rate constant for Cs-bo; DR=dynamic ratio; Ol%=percentage of open land in the catchment area; For%=percentage of forest land in the catchment area; Br = the relief of the catchment area.

Equation (4.3) should only be used when For%, Ol%, DR and Br are within the following ranges:

For%	Ol%	DR	Br
62-96	0-20	0.06-0.24	34-194

The most important parameter to explain variations in kbo is thus the lake's dynamic ratio. The higher the DR-value, the greater the resuspension activity, the smaller kbo and the slower the recovery. The second most important parameter comes from the catchment area. The greater the percentage of forest land (For%) and the smaller the percentage of open land (Ol%), the less the runoff from soil to water of caesium, the smaller the secondary dose during the years after Chernobyl, and the greater the value of the rate constant. This seems understandable. The relief also influences the runoff; the larger the Br, the faster the runoff from land to water.

Figure 4.9 gives a comparison that shows the co-variation between empirical data on kbo and the model-calculated values according to Eq. (4.3). The degree of explanation is 80%. In Table 4.6, a summary is given of important information on Eq. (4.3). This concerns how Eq. (4.3) can or <u>cannot</u> be used for our other lakes. Here, all lakes marked with a cross are those where Eq. (4.3) cannot be used. This is also shown in a column where model data on kbo have been prepared. In as many as 9 cases we get negative kbo-values.

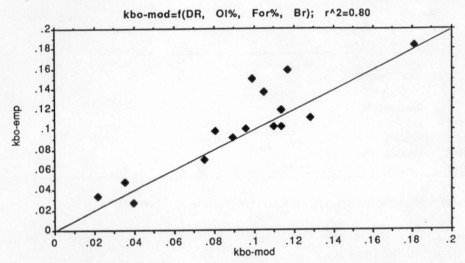

Fig. 4.9. Comparison between empirical data (y-axis) on kbo and model-calculated data (according to Eq.4.3) on the x-axis. The model gives an r^2-value of 0.80 (n=15).

This is for lakes that are <u>outside the range</u> of the model and implies that if the model is used for these lakes, then it will not predict a decrease in Cs-bo with time but an increase. This is a typical weakness with correlative ecometric models. It is also a weakness found in other models as well, but nonetheless it must be emphasized here that time-dependent ecometric models can only be used for lakes that fulfil the fundamental requirements of the model concerning the range of the parameters included.

Table 4.6. Information on Eq. (4.3). Data on the model parameters (For%, Ol%, DR and Br). Lakes outside the range of the model are marked. Comparison between empirical data and model data on kbo.

Lake	For%	Ol%	DR	Br	Outside range for kbo	Outside range for Cs-bo86	kbo-mod	kbo-emp
2101	67,6	16,0	0,24	37,4	x		-0,037	
2102	86,6	2,2	0,54	20,3	x	x	-0,051	
2103	78,0	9,0	0,36	14,3	x	x	-0,063	
2104	79,4	1,8	0,42	21,7	x	x	-0,033	
2105	68,0	0,0	0,14	131,2			0,112	
2106	100,0	0,0	0,07	38,5	x		0,167	
2107	79,0	5,0	0,27	73,0	x	x	0,028	
2108	78,5	0,4	0,25	34,1	x		0,030	
2109	91,4	0,2	0,15	44,4			0,096	
2110	86,4	0,5	0,14	33,8			0,088	0,092
2111	77,0	0,3	0,32	43,8	x	x	0,015	
2112	70,5	1,3	0,28	36,5	x		0,011	
2113	67,3	0,2	0,61	8,7	x	x	-0,099	
2114	70,6	6,4	0,49	25,8	x	x	-0,069	
2115	95,1	0,0	0,18	25,1		x	0,068	
2116	64,6	2,1	0,28	24,8	x	x	-0,010	
2117	62,5	0,9	0,18	34,0			0,041	0,027
2118	73,9	0,0	0,56	32,9	x	xx	-0,044	
2119	93,1	0,0	0,14	60,1			0,114	0,119
2120	96,0	0,0	0,19	44,4			0,081	0,098
2121	75,9	9,3	0,14	193,7			0,094	0,100
2122	83,2	0,0	0,14	83,6			0,114	0,103
2201	89,3	0,3	0,06	55,8			0,180	/0.184/
2202	74,9	0,0	0,10	23,6		xx	0,097	
2203	84,0	0,0	0,13	42,0		x	0,100	
2204	82,1	3,5	0,13	29,0		x	0,073	
2205	90,7	4,0	0,15	46,3			0,081	
2206	77,4	0,9	0,12	46,0			0,100	0,150
2207	64,4	0,0	0,29	51,7	x	x	0,017	
2208	78,9	1,9	0,19	22,4		x	0,036	
2209	79,3	1,5	0,10	35,2			0,107	
2210	67,0	0,2	0,20	15,8		x	0,016	
2211	67,1	1,9	0,27	22,8	x	x	-0,006	
2212	78,7	0,5	0,11	53,2			0,115	0,159
2213	86,1	2,9	0,13	80,1			0,110	0,103
2214	84,1	1,3	0,09	42,3			0,127	0,112
2215	82,9	7,4	0,11	79,9			0,103	0,137
2216	78,5	11,0	0,10	45,4			0,076	0,070
2217	73,3	19,7	0,11	61,7			0,037	0,048
2218	82,8	7,6	0,24	53,7			0,023	0,034
Max	96,0	19,7	0,24	193,7				
Min	62,5	0,0	0,06	33,8				

Step 3 — Cs-bo86

An interesting question is now whether Eq. (4.2) can be based on general, "constant" environmental parameters. We must then find out if Cs-bo86, as in the case of kbo, can be expressed in terms of general parameters. The initial tests cover the entire data material from the 15 lakes, thus also including the extreme value from lake 2201. But in this context, where the variation between lakes in Cs-bo86 is to be discussed, this extreme value hides more of the true relationships than it reveals. Figure 4.10A illustrates the relationship between Cs-soil and empirical data on Cs-bo86. Here we can clearly see that the value from lake 2201 is an extreme one. This can be seen even better in Fig. 4.10B which shows the best model that has been obtained (using step-wise multiple regression) when the Cs-bo86 value from lake 2201 is included in the material. We can then obtain an equation that is based on the area of A-bottoms (BA), the percentage of mires in the catchment area (Mire%) and the primary dose (Cs-soil) which gives an 87% degree of explanation of the variation in Cs-bo86. This is a very high degree of explanation for only three parameters.

> However, it is also a classical situation where the high r^2-value depends on an uneven distribution — one single high value (from lake 2201) causes an extremely strong imbalance in the regression. Consequently, the Cs-bo86 value from lake 2201 will be omitted in the following.

Subsequently, we will start from the ratio Cs-bo86/Cs-soil as a function of "constant" environmental factors since Cs-bo86 ought to follow Cs-soil — if the fallout is zero then also Cs-bo86 should be zero. In Table 4.7, showing the r-ranking, we can see that the ratio Cs-bo86/Cs-soil shows a significant co-variation with the theoretical water turnover time (T; r=0.62) and with the percentage of open land in the catchment area (Ol%; r=-0.55). Step-wise multiple regression (Table 4.8) shows that we can get a 77% degree of explanation with a model that gives log(Cs-bo86) as a function of:

Table 4.7. Table showing r-ranking for the ratio Cs-bo86/Cs-soil.

Parameter	Cs-bo86/Cs-soil	Parameter	Cs-bo86/Cs-soil
Cs-bo86/Cs-soil	1.00	BA	-0.02
T	0.62	Acid%	-0.03
Mor%	0.37	Q	-0.14
Mire%	0.35	ADr	-0.17
Dm	0.34	Rock%	-0.19
area	0.32	DR	-0.23
Vol	0.31	Ol%	-0.55
Lake%	0.12		
Br	0.07		
Im%	0.01		
For%	0.00		

(1) the theoretical water turnover time (T), the slower the turnover time, the greater the T and the greater value of Cs-bo86;

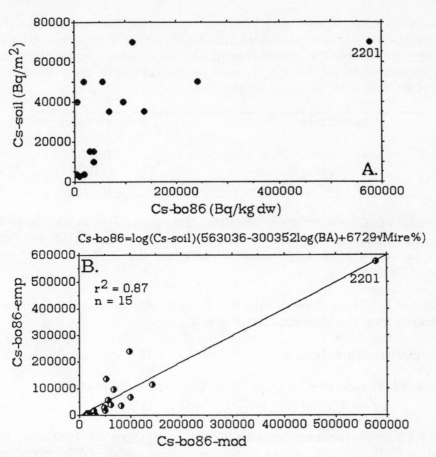

$Cs\text{-}bo86 = log(Cs\text{-}soil)(563036 - 300352log(BA) + 6729\sqrt{Mire\%})$

Fig. 4.10. A. Relationship between Cs-soil and Cs-bo86 with special marking of the "outlier", lake 2201.
B. Comparison between empirical data (y-axis) on caesium in sediment traps close to the bottom and model-calculated data (according to the given equation, which gives an r^2-value of 0.87) on the x-axis. The "outlier", lake 2201, has been marked.

(2) the percentage of open land in the catchment area (Ol%); the higher the Ol%, the lower the value of Cs-bo86; the higher the retention of caesium in the soil;

(3) the percentage of lakes in the catchment area (Lake%); the larger the Lake%, the larger the value of Cs-bo86. This may initially appear confusing since the lakes upstream function as sinks for caesium, which should lead to less transport of caesium to lakes downstream. This need not, however, apply to the particular situation in 1986. Both the water turnover and the caesium turnover in the lakes are much faster than the caesium turnover in soils. This means that it is logical that the Lake% works in the opposite direction to Ol% in this context.

Table 4.8. Results from step-wise multiple regression analysis. Parameters that statistically explain the variation in log(Cs-bo86)/Cs-soil) between the lakes. The model that yields the highest degree of explanation for one lake parameter (T) and three catchment area parameters (Ol%, Lake% and Br) is given as Eq. (4.4).

	Parameter	**r^2-value**
Step 1	$\sqrt{T}$	0.39
Step 2	$\sqrt{T}$, $\sqrt{Ol\%}$	0.57
Step 3	$\sqrt{T}$, $\sqrt{Ol\%}$, $\sqrt{Lake\%}$	0.65
Step 3	$\sqrt{T}$, $\sqrt{Ol\%}$, $\sqrt{Lake\%}$, $\sqrt{Br}$	0.77

(4) the relief of the catchment area (Br); this parameter is also included in this context, as in Eq. (4.3) for kbo; relief appears to be of great importance for runoff from land to water. The larger the Br-value the higher the value of Cs-bo86.

(5) Cs-soil has been "forced" into this least-square adaptation (it, of course, originates from the ratio Cs-bo86/Cs-soil).

The equation can be written:

$$\log(\text{Cs-bo86}) = \text{Cs-soil} \cdot (-1.928 + 2.632\sqrt{T} - 0.687\sqrt{Ol\%} + 0.865\sqrt{Lake\%} + 0.348\sqrt{Br}) \quad (4.4)$$
$$(r^2 = 0.77, \ n = 14)$$

where Cs-bo86=caesium in bottom sediment traps in 1986 (Bq/kg dry weight); T=theoretical water turnover time (years); Ol%=percentage of open land in the catchment area; Lake%=percentage of lakes in the catchment area.

Equation (4.4) applies when:

Cs-soil	T	Ol%	Lake%	Br
3000-70 000	0.02-1.6	0-19.7	0-6.8	33.8-193.7

The relationship between the empirical data and the data predicted by the model is given in Fig. 4.11. With this model, the values of Cs-bo86 can be predicted for other lakes. This is done in Table 4.9, where a list is given of the parameters included, their range and an explanation of how model 4.4 can be used for our 41 lakes. **As many as 17 of these lakes fall outside the rele-vant ranges.** The model should not be used for these lakes. The predictions are given in Table 4.10, which is divided into two parts:

101

Table 4.9. Information on Eq. (4.4). Data on the parameters included (Cs-soil, T, Ol%, Lake% and Br). Comparison between empirical data and model-calculated data and information on the lakes that fall outside the range of the model.

Lake	Cs-soil (Bq/m^2)	T (yr)	Ol%	Lake%	Br	Cs-bo86-emp (Bq/kg dw)	Cs-bo86-mod (Bq/kg dw)	Outside range
2101	15000	0,89	16	0	37,4		-986	
2102	4000	0,02	2,2	4,8	20,3		3553	x
2103	10000	1,19	9	0	14,3		1971	x
2104	7500	0,35	1,8	2,8	21,7		13315	x
2105	4000	0,7	0	9	131,2		27417	
2106	4000	0,42	0	0	38,5		7745	
2107	15000	1,74	5	1	73		57670	x
2108	4000	0,94	0,4	1,1	34,1		12511	
2109	4000	0,14	0,2	1,1	44,4		7901	
2110	3000	0,48	0,5	1,3	33,8	8400	7255	
2111	3000	0,09	0,3	10,4	43,8		10733	x
2112	4000	0,46	1,3	1,3	36,5		8647	
2113	10000	0,12	0,2	5,7	8,7		17678	x
2114	7500	0,03	6,4	4,6	25,8		3094	x
2115	4000	0,26	0	0	25,1		4628	x
2116	2500	0,34	2,1	5,4	24,8		5884	x
2117	10000	0,53	0,9	3,7	34	36700	30287	
2118	10000	0,06	0	8,5	32,9		32344	xx
2119	4000	1,2	0	3,4	60,1	19200	20988	
2120	4000	0,3	0	0	44,4	3300	7328	
2121	15000	0,47	9,3	0	193,7	37700	39360	
2122	15000	0,38	0	0	83,6	29800	43136	
2201	70000	0,33	0,3	0,8	55,8	/576000/	180623	
2202	40000	2,2	0	4	23,6		215799	xx
2203	50000	2,9	0	3,7	42		323580	x
2204	35000	0,65	3,5	1,5	29		64448	x
2205	35000	1,1	4	0,8	46,3		90966	
2206	35000	0,48	0,9	1,1	46	68100	87869	
2207	50000	2,4	0	0,3	51,7		256196	x
2208	50000	1,1	1,9	2,2	22,4		140725	x
2209	40000	0,85	1,5	0,5	35,2		93303	
2210	50000	0,68	0,2	3,3	15,8		144448	x
2211	50000	1,6	1,9	3,2	22,8		183103	x
2212	50000	0,24	0,5	4,7	53,2	241000	164433	
2213	35000	1,6	2,9	0	80,1	137000	117061	
2214	70000	0,02	1,3	1,3	42,3	115000	63726	
2215	40000	0,44	7,4	0,4	79,9	95100	64244	
2216	50000	0,37	11	6,8	45,4	56200	99715	
2217	50000	0,21	19,7	3,8	61,7	19100	32408	
2218	40000	0,1	7,6	1,1	53,7	6400	18698	
Max	70000	1,6	19,7	6,8	193,7	241000		
Min	3000	0,02	0	0	33,8	3300		

Table 4.10. Upper. Model data on caesium in material from bottom sediment traps from 1986 to 2030 without correction for the delay factor and the natural radioactive decay.
Lower. Model data for the period 1986 to 2030 with correction for the delay factor (1.5) and with consideration of natural radioactive decay.

Model data on Cs-bo (Bq/kg dw) - no decay, no delay

Lake	kbo-mod	1986	1987	1988	Year 1990	1995	2000	2010	2030
2105	0,112	27417	4575	1194	81	0	0		0
2109	0,096	7901	1699	537	54	0	0	0	0
2110	0,088	7255	1784	623	76	0	0	0	0
2117	0,041	30287	15838	9740	3683	324	28	0	0
2119	0,114	20988	3408	872	57	0	0	0	0
2120	0,081	7328	2008	760	109	1	0	0	0
2121	0,094	39360	8707	2809	292	1	0	0	0
2122	0,114	43136	7005	1792	117	0	0	0	0
2201	0,180	180623	10141	1170	16	0	0	0	0
2205	0,081	90966	24729	9310	1320	10	0	0	0
2206	0,100	87869	17827	5389	493	1	0	0	0
2209	0,107	93303	16839	4662	357	1	0	0	0
2212	0,115	164433	26321	6661	427	0	0	0	0
2213	0,110	117061	20120	5371	383	1	0	0	0
2214	0,127	63726	8317	1806	85	0	0	0	0
2215	0,103	64244	12304	3562	299	1	0	0	0
2216	0,076	99715	29634	11928	1932	20	0	0	0
2217	0,037	32408	18000	11581	4794	528	58	1	0
2218	0,023	18698	12998	9895	5735	1467	375	25	0

Model data on Cs-bo (Bq/kg dw) - with decay (e^(-0.023t)) and delay of 1.5 from 1990.

Lake	1986	1987	1988	1990	1995	2000	2010	2030
2105	27417	4369	1115	111	0	0	0	0
2109	7901	1623	501	73	0	0	0	0
2110	7255	1704	582	104	0	0	0	0
2117	30287	15126	9090	5039	395	31	0	0
2119	20988	3255	814	78	0	0	0	0
2120	7328	1917	710	149	1	0	0	0
2121	39360	8316	2621	400	1	0	0	0
2122	43136	6690	1673	160	0	0	0	0
2201	180623	9685	1092	21	0	0	0	0
2205	90966	23617	8689	1805	12	0	0	0
2206	87869	17026	5030	674	2	0	0	0
2209	93303	16082	4352	489	1	0	0	0
2212	164433	25137	6217	584	1	0	0	0
2213	117061	19215	5013	524	1	0	0	0
2214	63726	7943	1685	116	0	0	0	0
2215	64244	11751	3325	408	1	0	0	0
2216	99715	28302	11133	2644	25	0	0	0
2217	32408	17191	10809	6559	645	63	1	0
2218	18698	12413	9236	7847	1789	408	21	0

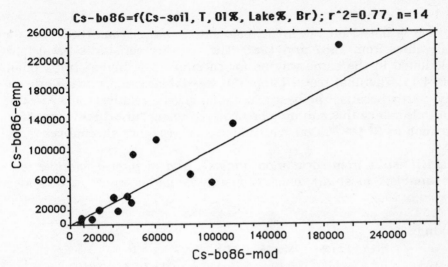

Fig. 4.11. Comparison between empirical data (y-axis) on Cs-bo86 and model-calculated data (according to Eq. 4.4) on the x-axis. The model gives an r^2-value of 0.77 (n=14).

• The upper part concerns model predictions for all lakes where Eqs. (4.3) and (4.4) can be used. This concerns 19 of our 41 lakes. The reason for there being so few is, naturally, that the model was only developed for 14 lakes. This gives it a restricted application. In Part 1 of the table only "straight" model values are given, i.e., values without corrections.

• The lower part gives Cs-bo values corrected for both the delay factor (1.5) and for natural radioactive decay (Cs-137 has a half-life of 30.2 years). The results in this part of the table will be used later.

Thus, Table 4.10 gives an initial prediction of how the recovery may appear for a dose parameter. This prediction suffers from several uncertainties. Naturally, the first concerns the fundamental problem of it only being based on data from 3 years. The other problem concerns the uncertainty with the delay factor. From Table 4.10, we can see that there are considerable differences in the probable development of this dose parameter in the lakes.

4.3.2. CAESIUM IN LAKE WATER

☞ **The task in this section is to discuss the future development for caesium in lake water.**

This may be calculated in several ways, e.g.,:

Method 1

This method is based on the model-calculated values from 1986 and the empirical values from 1987 and 1988. The recovery constants (kwa) have been calculated in the same way as for caesium in sediment traps (kbo), see Table 4.11. The final result (Step 3 in this table) is not particularly satisfactory, partly because the best adaptation gives a relatively low r^2-value (0.66), but also since this can only be achieved using **"unethical" transformations such as $2^{Lake\%}$**. Consequently, we will try other alternatives.

Table 4.11. Results from correlation analysis and step-wise multiple regression analysis to study which "constant" environmental factors influence kwa.

A. r-ranking

Dm	Vol	area	For%	Mor%	ADr	Q	Mire%	Rock%
0.62	0.38	0.27	0.14	0.10	0.04	0.03	0.03	0.02

Br	Lake%	Ol%	BA	DR
-0.07	-0.17	-0.18	-0.30	-0.42

B. Step-wise multiple regression

	Parameter	r^2-value
Step 1	log(Dm)	0.44
Step 2	log(Dm), Vol^2	0.54
Step 3	log(Dm), Vol^2, $2^{Lake\%}$	0.66

Method 2

This method uses Eq. (4.1) as startingpoint, which gives Cs-wa(t) as a function of Cs-bo(t), Q, Dm and Br. The question is: Can this model be used to predict Cs-wa? Since the model is of the type: Cs-wa(t) = Cs-bo(t) + a constant, we may expect that Cs-wa(t) moves towards this constant when Cs-bo(t) moves towards zero. Table 4.12 shows that this is the case. This type of model **should not**, consequently, be used to predict the future development of Cs-wa.

Method 3

This method is based on the empirical material for Cs-bo from 1987 and 1988 and on the assumption that a relationship can be obtained between the ratio Cs-wa/Cs-bo and "constant" environmental factors. This also implies that Cs-wa will move towards zero when Cs-bo does likewise. The question is what factors influence this ratio? We have tested absolute and logarithmic values. The best results have been obtained with logarithmic data. From Table 4.13, we can further note that particularly the propor-

Table 4.12. Model data on caesium in lake water from 1986 to 2030 with correction for the delay factor (1.5) and with consideration of natural radioactive decay.

Model data on Cs-wa (Bq/l) - with radioactive decay ($e^{(-0.023t)}$) and a delay of 1.5 from 1990.

Year

Lake	1986	1987	1988	1990	1995	2000	2010	2030
2105	0,668	0,167	0,097	0,075	0,073	0,073	0,073	0,073
2109	0,234	0,098	0,074	0,064	0,063	0,063	0,063	0,063
2110	0,235	0,115	0,090	0,080	0,078	0,078	0,078	0,078
2117	0,727	0,398	0,267	0,179	0,079	0,071	0,070	0,070
2119	0,464	0,079	0,026	0,010	0,008	0,008	0,008	0,008
2120	0,130	0,013	-0,014	-0,026	-0,029	-0,029	-0,029	-0,029
2121	0,951	0,277	0,154	0,105	0,097	0,097	0,097	0,097
2122	0,978	0,187	0,078	0,045	0,042	0,042	0,042	0,042
2201	4,331	0,621	0,435	0,411	0,411	0,411	0,411	0,411
2205	2,274	0,812	0,488	0,339	0,300	0,300	0,300	0,300
2206	2,232	0,695	0,435	0,340	0,325	0,325	0,325	0,325
2209	2,372	0,696	0,442	0,358	0,347	0,347	0,347	0,347
2212	3,732	0,709	0,298	0,176	0,164	0,163	0,163	0,163
2213	2,671	0,548	0,240	0,143	0,131	0,131	0,131	0,131
2214	1,752	0,542	0,406	0,372	0,370	0,370	0,370	0,370
2215	1,506	0,367	0,184	0,121	0,112	0,112	0,112	0,112
2216	2,201	0,652	0,279	0,095	0,038	0,038	0,038	0,038
2217	0,780	0,449	0,311	0,219	0,090	0,078	0,076	0,076
2218	0,463	0,327	0,258	0,228	0,096	0,066	0,058	0,057

Table 4.13. r-ranking table for the ratio Cs-wa/Cs-bo.

Cs-wa/Cs-bo		log(Cs-wa)/log(Cs-bo)	
Cs-wa/Cs-bo	1,00	log(Cs-wa)/log(Cs-bo)	1,00
Br	0,36	Im%	0,77
Mire%	0,34	Rock%	0,52
Mor%	0,20	Dm	0,45
Acid%	0,16	Q	0,40
DR	0,09	ADr	0,39
BA	0,05	Lake%	0,23
T	0,01	Vol	0,18
Vol	0,01	Mire%	0,07
A	0,00	T	0,06
For%	-0,11	Ol%	0,03
Q	-0,11	Br	0,01
log(Cs-wa)/log(Cs-bo)	-0,12	A	-0,05
Dm	-0,13	Cs-wa/Cs-bo	-0,12
ADr	-0,13	For%	-0,14
Rock%	-0,16	Mor%	-0,38
Im%	-0,17	BA	-0,47
Lake%	-0,19	DR	-0,61
Ol%	-0,23	Acid%	-0,77

tion of acidic bedrocks in the catchment area (Acid%) and the dynamic ratio of the lakes (DR) have high r-values in relation to the ratio with logarithmic values. In the subsequent "ladder", we can see the results from the step-wise multiple regression:

	Parameters		**r^2-value**
Step 1	$\sqrt{Acid\%}$		0.61
Step 2	$\sqrt{Acid\%}$,	$10^{(BA-90)}$	0.75
Step 3	$\sqrt{Acid\%}$,	$10^{(BA-90)}$, Vol	0.81
Step 4	$\sqrt{Acid\%}$,	$10^{(BA-90)}$, Vol, log(Mire%)	0.83

With four "constant" environmental parameters, we can thus achieve an 83% degree of explanation.

The factor $10^{(BA-90)}$ is **not** as difficult to interpret as might appear. In all lakes, there is a shoreline zone with erosion and transportation bottoms and in lakes larger than 1 km^2 this zone is rarely less than 15% of the lake area (see Håkanson and Jansson, 1983). In smaller lakes, such as ours, up to 90% of the bottom area may be made up of A-bottoms. This is the explanation of (BA-90). The best regression is given by:

$$\log(Cs\text{-}wa)/\log(Cs\text{-}bo) = 29.024\text{Vol} - 1212.323 \cdot 10^{(BA-90)} - 0.023\sqrt{Acid\%} + 0.082\log(Mire\%) - 0.183$$
$$(r^2=0.83, \ n=30) \tag{4.5}$$

where Vol=lake volume in km^3; BA=area of accumulation bottoms (%); Acid%=the area of acidic bedrocks in the catchment area (%); Mire%=the area of mires in the catchment area (%).

Equation (4.5) applies for:

Cs-bo	Vol	BA	Acid%	Mire%
311-23 100	0.0002-0.004	35.3-86.4	0-100	3-21

The lake volume influences the water turnover time, BA influences the resuspension conditions, Acid% probably influences the soil chemistry so that more caesium is present in lake water if Acid%, and thus water pH, decrease, and mires influence the runoff of both acidic substances and caesium.

If we use the relationships given by Eqs (4.5) and (4.2), which give Cs-bo as a function of time, we can calculate how Cs-wa should vary during the immediate future in our 15 lakes. The results are shown in Table 4.14. As earlier for Cs-bo, a "delay factor" of 1.5 has been used and the data have been corrected for natural radioactive decay. But in this case, the model gives **completely incorrect** results. This depends on the model being of the type: Cs-wa=Cs-bo^{const}. The constant is given by Eq. (4.5), but since it has, for all lakes, a smaller absolute value than -1 (e.g., -0.56 for lake 2110), Cs-wa will increase when Cs-bo decreases, as shown in Table 4.14. Thus, Eq. (4.5) can only be used to describe the conditions during the two actual years.

Table 4.14. Model-calculated time development (1986-2030) according to Eq. (4.5) for caesium in lake water (Cs-wa).

Lake	1986 Cs-wa-mod (Bq/l)	1987 Cs-wa-emp	1988 Cs-wa-emp	1990	1995	2000	2010	2030
2110	0,26	0,04	0,01	-1,56	0,07	1,51	31,53	13720,00
2117	0,87	0,12	0,13	0,05	0,06	0,14	0,33	1,72
2119	0,42	0,13	0,10	0,09	0,21	2,54	31,23	4720,73
2120	0,04	0,03	0,07	0,10	0,17	0,99	5,61	180,83
2121	0,91	0,14	0,14	0,09	0,16	0,93	5,54	194,15
2122	0,69	0,16	0,17	0,25	0,39	1,41	5,15	68,39
2201	12,90	0,79	0,66	0,52	0,75	2,10	5,88	46,14
2206	1,80	0,51	0,28	0,40	0,50	0,96	1,84	6,84
2212	5,39	0,50	0,40	0,50	0,61	1,05	1,81	5,41
2213	3,10	0,64	0,43	0,71	0,78	1,02	1,33	2,26
2214	2,86	0,64	0,64	0,53	0,66	1,29	2,52	9,52
2215	2,17	0,15	0,09	0,10	0,18	1,09	6,76	259,94
2216	1,26	0,61	0,29	0,23	0,29	0,61	1,25	5,38
2217	0,49	0,18	0,07	0,20	0,21	0,32	0,48	1,09
2218	0,20	0,07	0,06	0,06	0,06	0,10	0,15	0,39

Neither has it been possible to obtain sufficiently good models for Cs-wa by means of other simple manipulations. This probably depends on several reasons; partly that the starting values of Cs-wa86 are calculated from a model which implies a certain degree of uncertainty, and partly that the variation in Cs-wa with time is probably so large that our empirical lake mean values for Cs-wa are not sufficiently good.

4.3.3. DECREASE IN PERCH

☞ **The aim of this exercise is to calculate the future development for caesium in perch fry.**

Figure 3.11A gave Cs-pe86 on the y-axis and Cs-pe87 on the x-axis. This figure clearly shows that the value of Cs-pe86 from lake 2211 of 524 000 Bq/kg ww is abnormally high. It has been omitted from the subsequent calculations in this section. Cs-pe86 is first calculated as a function of "constant" environmental parameters. The model giving the highest degree of explanation is:

$$\log(\text{Cs-pe86}) = 0.112 - 0.0340\text{I\%} + 0.412\sqrt{T} - 0.004\text{Dm}^2 + 0.942\log(\text{Cs-soil}) + 0.00001\text{Br}^2 \quad (4.6)$$
$$(r^2 = 0.87, \ n=39)$$

where Ol%=Percentage of open land; T=theoretical water turnover time (years); Dm=mean depth (m); Cs-soil=fallout of Cs (Bq/m^2); Br=relief of the catchment area.

Equation (4.6) applies when (see also Table 4.15, which gives a comparison between empirical data and model data from Eq. 4.6):

Ol%	T	Dm	Cs-soil	Br
0-20	0.02-2.9	1.1-10.1	2500-70 000	9-131

The steps are:

	Parameters	r^2-value
Step 1	log(Cs-soil)	0.68
Step 2	log(Cs-soil), $\sqrt{T}$	0.77
Step 3	log(Cs-soil), $\sqrt{T}$, Ol%	0.83
Step 4	log(Cs-soil), $\sqrt{T}$, Ol%, Br^2	0.85
Step 5	log(Cs-soil) $\sqrt{T}$, Ol%, Br^2, Dm^2	0.87

The relationship between empirical data on log(Cs-pe86) and model data (according to Eq. 4.6) is shown in Fig. 4.12A; the r^2-value is 0.87. In Fig. 4.12B, it is emphasized that this does not mean that the degree of explanation between the absolute values for Cs-pe86 is also 87%. **In this case, the r^2-value decreases from 0.87 to 0.68 when use is not made of logarithmic values.** We can also see that one lake (2207) falls outside the frame as regards the model-predicted values. If we remove this lake, then a much better relationship is obtained between empirical data and model data according to Eq. (4.6). The r^2-value will then be 0.87 (Fig. 4.12C).

Table 4.16 gives the r-ranking test of which factors influence the variation in Cs-pe between the lakes for all the data from 1986, 1987 and 1988. Since the dose parameters (Cs-wa and Cs-bo) are included in this test, we only have data from 15 lakes. The water chemical parameters are mean values from the production period of each year. For simplicity, in this section (i.e., 4.4.3), these values are not denoted as, e.g., pH-86-5/9 but only pH. Table 4.16 shows which factors have large or small, positive or negative links to either Cs-pe or log(Cs-pe).

The results from the step-wise multiple regression analysis are shown, first for caesium in perch in relation to caesium in water and then the relation to caesium in bottom traps.

	Parameters	r^2-value
Step 1	log(Cs-wa)	0.77
Step 2	log(Cs-wa), $temp^2$	0.82
Step 3	log(Cs-wa), $temp^2$, $totP^2$	0.85

Table 4.15. Information on Eq. (4.6). Included parameters (Cs-soil, Ol%, T, Dm and Br). Comparison between empirical data and model data on Cs-pe86.

Lake	Cs-soil (Bq/m^2)	Ol%	T (yr)	Dm (m)	Br	Cs-pe86-emp (Bq/kg ww)	Cs-pe86-mod (Bq/kg ww)
2101	15000	16	0,89	4,0	37,4	12900	6917
2102	4000	2	0,02	1,1	20,3	2800	3122
2103	10000	9	1,19	2,3	14,3	9900	10098
2104	7500	2	0,35	3,1	21,7	11300	8019
2105	4000	0	0,70	2,7	131,2	7200	9656
2106	4000	0	0,42	4,7	38,5	3390	4988
2107	15000	5	1,74	4,3	73,0	33400	24900
2108	4000	0	0,94	3,3	34,1	5600	7449
2109	4000	0	0,14	3,2	44,4	8200	4336
2110	3000	0	0,48	4,7	33,8	2300	3939
2111	3000	0	0,09	2,1	43,8	1730	3249
2112	4000	1	0,46	1,7	36,5	7160	5646
2113	10000	0	0,12	1,2	8,7	13300	10416
2114	7500	6	0,03	1,2	25,8	2200	4268
2115	4000	0	0,26	2,0	25,1	4600	5073
2116	2500	2	0,34	3,5	24,8	4200	2767
2117	10000	1	0,53	4,2	34,0	16900	12201
2118	10000	0	0,06	1,1	32,9	15800	9692
2119	4000	0	1,20	2,7	60,1	12700	9157
2120	4000	0	0,30	2,1	44,4	5200	5395
2121	15000	9	0,47	2,4	193,7	25700	22753
2122	15000	0	0,38	2,8	83,6	27000	21634
2201	70000	0	0,33	8,3	55,8	51800	46439
2202	40000	0	2,20	6,2	23,6	81200	81243
2203	50000	0	2,90	10,1	42,0	66616	70603
2204	35000	4	0,65	8,5	29,0	32768	20309
2205	35000	4	1,10	6,9	46,3	17577	33008
2206	35000	1	0,48	6,4	46,0	23500	31644
2207	50000	0	2,40	4,9	51,7	54334	127686
2208	50000	2	1,10	6,0	22,4	74299	57986
2209	40000	2	0,85	8,1	35,2	37400	32237
2210	50000	0	0,68	6,2	15,8	•	53308
2211	50000	2	1,60	6,1	22,8	•	70419
2212	50000	0	0,24	4,9	53,2	76200	46904
2213	35000	3	1,60	5,1	80,1	96000	58721
2214	70000	1	0,02	2,9	42,3	49510	48280
2215	40000	7	0,44	4,5	79,9	23800	28994
2216	50000	11	0,37	4,0	45,4	44794	23480
2217	50000	20	0,21	3,8	61,7	6400	10611
2218	40000	8	0,10	2,1	53,7	9100	20667
2219	35000	0	1,10	4,6	19,7	58030	55422

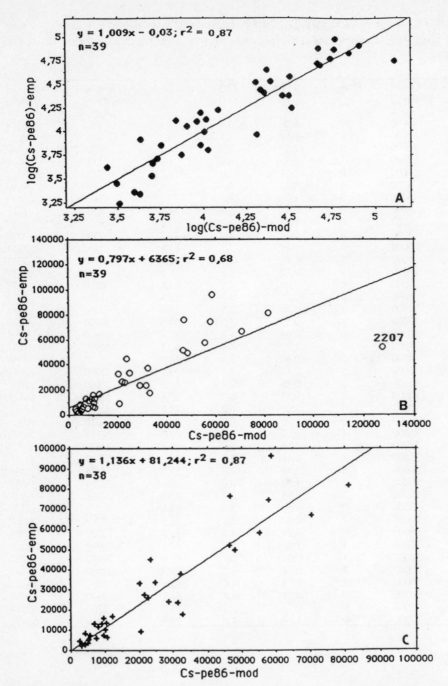

Fig. 4.12. Relationship between empirical data on caesium in 1+perch (y-axis) and model-calculated data (x-axis).

A. According to Eq. (4.6), i.e., for logarithmic data, n=39. Degree of explanation=87%.

B. According to Eq. (4.6) but for absolute values on Cs-pe, n=39. Degree of explanation= 68%. The "model outlier", lake 2207 has been marked.

C. According to Eq. (4.6) for absolute values on Cs-pe, but with lake 2207 omitted. Degree of explanation=87%. Values in Bq/kg wet weight.

Table 4.16. r-ranking table for Cs-pe and log(Cs-pe); data from 1986, 1987 and 1988 from the 15 best investigated lakes.

Parameter	Cs-pe	Parameter	Cs-pe	Parameter	log(Cs-pe)	Parameter	log(Cs-pe)
Cs-pe	1,00	Br	0,03	log(Cs-pe)	1,00	T	0,09
log(Cs-pe)	0,82	For%	0,01	log(Cs-wa)	0,88	Fe	0,04
log(Cs-wa)	0,79	Col	-0,01	Cs-pe	0,82	area	-0,02
log(Cs-bo)	0,75	Fe	-0,12	log(Cs-bo)	0,75	Ol%	-0,10
Cs-wa	0,62	Mor%	-0,13	log(Cs-soil)	0,59	totP	-0,13
Cs-bo	0,61	totP	-0,15	Im%	0,54	For%	-0,15
Im%	0,49	Ol%	-0,15	Cs-soil	0,54	Mor%	-0,29
log(Cs-soil)	0,44	BA	-0,28	Cs-wa	0,48	BA	-0,34
Cs-soil	0,44	cond	-0,35	Cs-bo	0,46	DR	-0,50
Dm	0,34	CaMg	-0,37	Rock%	0,41	cond	-0,52
Rock%	0,33	Ca	-0,40	Dm	0,36	Acid%	-0,55
Q	0,21	Tr	-0,41	Q	0,27	pH	-0,55
ADr	0,20	DR	-0,41	ADr	0,26	CaMg	-0,55
Vol.	0,16	alk	-0,41	Mire%	0,20	Tr	-0,59
T	0,14	pH	-0,44	Br	0,16	Ca	-0,59
Lake%	0,10	Acid%	-0,50	Col	0,15	alk	-0,63
Mire%	0,08			Vol.	0,15		
area	0,03			Lake%	0,13		

The best least-square adaptation for the three parameters these three years is given by:

$$\log(\text{Cs-pe}) = 4.929 + 0.704\log(\text{Cs-wa}) - 0.002\ \text{temp}^2 - 0.022\text{totP}^2 \qquad (4.7)$$
$$(r^2=0.85,\ n=45)$$

where Cs-pe=lake annual mean of Cs in 1+perch (Bq/kg ww); Cs-wa=mean value from the production period (May-Sept.) of Cs in lake water (Bq/l); temp=mean value from the production period of the lake water temperature (degrees C); totP=mean value (in µg/l) from the prodduction period.

Equation (4.7) applies when:

Cs-wa	temp	totP
0.01-0.79	9.9-20.1	7.1-27.8

The relationship with three steps thus gives an 85% degree of explanation.

When caesium in bottom sediment traps is used as dose parameter, we get a slightly poorer result.

	Parameters	r^2-value
Step 1	log(Cs-bo)	0.57
Step 2	log(Cs-bo), alk	0.70
Step 3	log(Cs-bo), alk, temp^2	0.78
Step 4	log(Cs-bo), alk, temp^2, Acid%	0.83

where Cs-bo=caesium in bottom sediment traps (Bq/kg dry weight);
alk=mean alkalinity from the production period (meq/l); Acid%=percentage of acidic bedrocks in the catchment area.

The equation is given by:

$$\log(\text{Cs-pe}) = 3.492 - 0.003\text{Acid}\% - 2.036\text{alk} - 0.002\text{temp}^2 + 0.322\log(\text{Cs-bo}) \quad (4.8)$$
$$(r^2=0.83, \ n=45)$$

The relationship applies when:

Cs-bo	Acid%	alk	temp
311-576 000	0-100	0.006-0.508	9.9-20.1

The relationship with these four parameters thus gives an 83% degree of
explanation.

In summary, we may note that Cs-pe can be described well ($r^2=0.85$) by
Eq. (4.7), which is based on Cs-wa (mean values from the production period) as dose parameter and water temperature and totP-concentration as
sensitivity parameters. The reason for including totP has already been
discussed. It is logical that water temperature is included, since it is of
great importance for bioproduction. The higher the temperature the higher, the bioproduction and the greater the "biological dilution" of a given
Cs-dose.

The recovery for caesium in 1+perch may, as with the recovery for Cs-bo,
probably best be described according to approach 2 (i.e., using an exponential function). If we then calculate the rate constants for all 15 lakes for
1+perch (kpe) **and attempt to link these kpe-values to "constant" environmental parameters**, the result is as shown in Table 4.17 as an r-ranking
table and as a "ladder". From this table, we can see that:

• a 78% explanatory model for the variation in kpe can be obtained;

• the size of the catchment area (ADr), its relief (Br), the water discharge
(Q), the primary dose (Cs-soil) and the area of A-bottoms (BA), in this
order, appear to influence kpe.

The best adaptation is given by:

$$\text{kpe} = 0.005 + 0.000152\text{BA} + 0.509\text{Q} - 0.006\text{ADr} - 0.0001239\text{Br} + 0.007\log(\text{Cs-soil}) \quad (4.9)$$
$$(r^2=0.78, \ n=15)$$

Table 4.17. Left: r-ranking for kpe.
Right: "Ladder" for kpe, i.e., results from step-wise multiple regression analysis to find out the factors of most importance in explaining kpe-variations between the lakes.

r-ranking	kpe		Parameters				r^2
kpe	1,00	Step 1	ADr				0,28
T	0,41	Step 2	ADr	Br			0,52
BA	0,37	Step 3	ADr	Br	Q		0,64
DR	0,21	Step 4	ADr	Br	Q og(Cs-soil)		0,73
For%	0,18	Step 5	ADr	Br	Q og(Cs-soil)	BA	0,78
Area	0,09						
Im%	0,07						
Lake%	0,06						
Lake	-0,05						
Ol%	-0,05						
Acid%	-0,06						
Vol.	-0,06						
Rock%	-0,07						
Mire%	-0,16						
Dm	-0,21						
Br	-0,22						
Cs-soil	-0,29						
Q	-0,51						
ADr	-0,53						

where BA=area of A-bottoms (% of lake area); Q=water discharge (m^3/s); Br=relief of the catchment area; Cs-soil=fallout of caesium (Bq/m^2).

Equation (4.9) applies when:

BA	Q	Br	Cs-soil
35.3-86.4	0.01-0.39	34-194	3000-70 000

Fig. 4.13 gives the relationship between empirical and model data on kpe.

Table 4.18A shows how the recovery can be predicted from Eq. (4.9) if we use empirical data on Cs-pe86 for all 15 lakes as the startingpoints. In this part of the table, no consideration has been taken to natural radioactive decay. This is done in Table 4.18B. The results in this part of the table will be used in the next section to predict the development in pike.

In summary, as regards the development of Cs-pe, we can see that Fig. 4.14 illustrates the recovery that can be predicted for the lake that will recover slowest (2201) and the lake that will make the fastest recovery (2120) according to Eq. (4.9). The ecometric model indicates that the Cs-concentration in 1+perch in lake 2201 should be below the guidelines issued by the National Swedish Food Administration (1500 Bq/kg ww) in about year 2000.

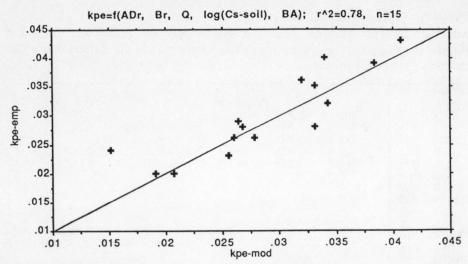

Fig. 4.13. Comparison between empirical data (y-axis) on kpe and mo-del-calculated data (according to Eq. 4.7) on the x-axis. The model gives an r^2-value of 0.78 (n=15).

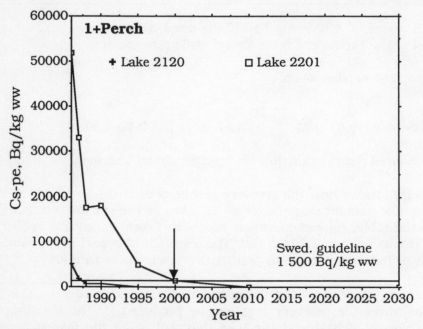

Fig. 4.14. Predicted recovery for caesium in 1+perch in the lake which, according to the model, will recover slowest (2201) and the lake predicted to recover fastest (2120) among our 15 lakes.

*Table 4.18. Prognosis of the development as regards caesium in 1+perch
up to 2030. Model data from 1990. Empirical data for 1986, 1987 and
1988.*
Upper. Model data without correction for natural radioactive decay.
Lower. Model data with correction for natural radioactive decay.

Cs in 1+perch (Bq/kg ww); no consideration to decay

Lake	kpe	Cs-pe86-emp	Cs-pe87-emp	Cs-pe88-emp	Cs-pe90-emp 1990	Cs-pe95 1995	Cs-pe00 2000	Cs-pe10 2010	Cs-pe30 2030
2110	0,029	2300	1502	451	572	100	18	1	0
2117	0,032	16900	10600	2857	3638	533	78	2	0
2119	0,028	12700	5500	2674	3312	617	115	4	0
2120	0,04	5200	1754	586	762	69	6	0	0
2121	0,02	25700	10618	8393	9840	2964	893	81	1
2122	0,035	27000	7951	3852	5032	616	75	1	0
2201	0,02	51800	33100	17588	19834	5974	1799	163	1
2212	0,028	76200	41600	16335	19873	3704	690	24	0
2213	0,043	96000	23400	8919	12187	923	70	0	0
2214	0,024	49510	30000	12839	15645	3707	878	49	0
2215	0,036	23800	20900	3250	4228	488	56	1	0
2216	0,039	44794	24570	5259	6890	664	64	1	0
2217	0,026	6400	4986	1487	1837	386	81	4	0
2218	0,026	9100	2707	2196	2612	549	115	5	0

Cs in 1+perch (Bq/kg ww); with consideration to decay

Lake	kpe	Cs-pe86-emp	Cs-pe87-emp	Cs.pe88-emp	Cs-pe90 1990	Cs-pe95 1995	Cs-pe00 2000	Cs-pe10 2010	Cs-pe30 2030
2110	0,029	2300	1502	451	521	82	13	0	0
2117	0,032	16900	10600	2857	3318	434	57	1	0
2119	0,028	12700	5500	2674	3021	502	83	2	0
2120	0,04	5200	1754	586	695	56	5	0	0
2121	0,02	25700	10618	8393	8975	2410	647	47	0
2122	0,035	27000	7951	3852	4590	501	55	1	0
2201	0,02	51800	33100	17588	18091	4857	1304	94	0
2212	0,028	76200	41600	16335	18126	3011	500	14	0
2213	0,043	96000	23400	8919	11116	751	51	0	0
2214	0,024	49510	30000	12839	14270	3014	636	28	0
2215	0,036	23800	20900	3250	3856	396	41	0	0
2216	0,039	44794	24570	5259	6284	540	46	0	0
2217	0,026	6400	4986	1487	1676	314	59	2	0
2218	0,026	9100	2707	2196	2383	446	84	3	0

4.3.4. DECREASE IN PIKE

☞ **The exercise here is to calculate the future development for caesium
in pike.** This is intriguing since the values still after 3-4 years <u>increase</u> in
most lakes. When will the peak be reached? What factors will influence the
"tail", i.e., the subsequent recovery?

The startingpoint here is that the Cs-concentration in pike follows the Cs-concentration in 1+perch but with a certain time lag. The question is first how different ratios between Cs-pi and Cs-pe can be linked to different dose and sensitivity parameters regulating this time lag. The correlation matrix (Table 4.19) shows a number of test results in order to find out the type of ratio that is most suitable to use. The time lag constant can be arranged in several ways, and below we have mostly tested variants of the type **Cs-pi/Cs-pe^x**, where the exponent x has been varied by values less than 1.

Which value of this delay constant (x) will then give the best adaptation to the available empirical material? Table 4.19 shows that the highest r^2-values with regard to the dose parameters Cs-bo and Cs-wa are obtained for the simplest ratio, namely Cs-pi/Cs-pe. However, if we were to use this ratio, we would get an unrealistically rapid recovery for Cs-pi. The recovery in Cs-pi will then follow the recovery in Cs-pe.

If we use a very small recovery constant, e.g., 0.01, the decrease in Cs-pi will be very slow. On the basis of existing empirical data for Cs-pi, it may be realistic to assume that the recovery constant for Cs-pi should be in the range 0.5-0.25. Here we will use the value 0.4. Future research will reveal whether this is realistic or if it is better to use a time-dependent exponent (like t*z=x). **The aim here has only been to demonstrate how this type of modelling can be done.** The choice of constant influences the linking to which dose and sensitivity parameters that are included in the model. The parameters used here (with the constant 0.4) will thus not necessarily be included in models based on other values of the constant.

Table 4.20 is an r-ranking table showing that the ratio Cs-pi/Cs-pe$^{0.4}$ can be linked mainly to the lake water temperature during the production period (temp), r=-0.54; i.e., the higher the temperature, the more caesium in pike in relation to perch (at a given time, t). As earlier for Cs-bo and Cs-pe, we must also consider Acid% (acidic bedrocks) and bottom dynamic conditions; the dynamic ratio DR gives a high r-value (-0.52).

Table 4.21 gives results from step-wise multiple regression analysis. We can see that a model giving a 79% degree of explanation can be obtained. The data then emanate from the 15 lakes for 1988 and 1987 for Cs-pi and from 1987 and 1986 for all other parameters; the pike were caught during the spring each year and the Cs-pi values are linked to the dose and sensitivity parameters from the year before capture. The model is based on:

Table 4.19. Correlation matrix for different ratios between caesium in pike and caesium in 1+perch and different dose parameters.

	Cs-pi(t)/Cs-pe(t)	log(Cs-pi(t)/Cs-pe(t))	Cs-pi(t)/√Cs-pe(t)	Cs-pi(t)/Cs-pe(t)^0.4	Cs-pi(t)/Cs-pe(t)^0.2	Cs-pi(t)/Cs-pe(t)^0.1	Cs-pi(t)/Cs-pe(t)^0.01	Cs-bo(t)	Cs-wa(t)	log(Cs-bo(t))	log(Cs-wa(t))
Cs-pi(t)/Cs-pe(t)	1,00										
log(Cs-pi(t)/Cs-pe(t))	0,96	1,00									
Cs-pi(t)/√Cs-pe(t)	0,67	0,67	1,00								
Cs-pi(t)/Cs-pe(t)^0.4	0,57	0,59	0,99	1,00							
Cs-pi(t)/Cs-pe(t)^0.2	0,41	0,44	0,95	0,98	1,00						
Cs-pi(t)/Cs-pe(t)^0.1	0,34	0,37	0,92	0,96	1,00	1,00					
Cs-pi(t)/Cs-pe(t)^0.01	0,29	0,31	0,88	0,93	0,99	1,00	1,00				
Cs-bo(t)	-0,32	-0,33	-0,06	0,00	0,11	0,16	0,20	1,00			
Cs-wa(t)	-0,31	-0,31	-0,03	0,04	0,15	0,19	0,23	1,00	1,00		
log(Cs-bo(t))	-0,59	-0,62	-0,13	-0,03	0,14	0,21	0,27	0,67	0,68	1,00	
log(Cs-wa(t))	-0,52	-0,51	0,08	0,18	0,34	0,40	0,44	0,70	0,72	0,88	1,00

Table 4.20. r-ranking table for the ratio Cs-pi(t)/Cs-pe-(t)$^{0.4}$.

	Cs-pi(t)/Cs-pe(t)^0.4		Cs-pi(t)/Cs-pe(t)^0.4		Cs-pi(t)/Cs-pe(t)^0.4
Cs-pi(t)/Cs-pe(t)^0.4	1,00	Vol.	0,14	Ol%	-0,13
Im%	0,52	Mire%	0,10	CaMg	-0,21
Q	0,31	Cs-wa	0,04	cond	-0,21
Dm	0,30	Forest%	0,02	alk	-0,22
ADr	0,30	Cs-bo	0,00	Ca	-0,23
Col	0,26	Area	-0,02	BA	-0,33
Rock%	0,26	Mor%	-0,09	totP	-0,35
Cs-pe	0,21	Fe	-0,10	DR	-0,42
T	0,15	Lake%	-0,12	Acid%	-0,52
Br	0,15	pH	-0,13	temp	-0,54

Table 4.21. Results from step-wise multiple regression for the ratio Cs-pi(t)/Cs-pe(t)0.4.

	Parameters					r^2-value
Step 1	Acid^0.25					0,38
Step 2	Acid^0.25	log(Cs-bo(t))				0,54
Step 3	Acid^0.25	log(Cs-bo(t))	BA			0,63
Step 4	Acid^0.25	log(Cs-bo(t))	BA	CaMg(t)^2		0,70
Step 5	Acid^0.25	log(Cs-bo(t))	BA	CaMg(t)^2	√Ol%	0,77
Step 6	Acid^0.25	log(Cs-bo(t))	BA	CaMg(t)^2	√Ol% Vol	0,79

- Acid% (most important); the higher the Acid%, the lower the ratio.
- Cs-bo, or rather log(Cs-bo).
- BA, i.e., the area of A-bottoms.
- Hardness of the lake water (CaMg).
- Percentage of open land in the catchment area (Ol%).
- Lake volume (Vol).

Of these six parameters Vol and Ol% have positive signs, i.e., the ratio Cs-pi/Cs-pe$^{0.4}$ increases when these parameters increase. The opposite applies to the other four parameters. The best least-square adaptation can be written:

Cs-pi(t)=Cs-pe(t)$^{0.4}$*(365 + 5241Vol - 1.27BA - 36.71log(Cs-bo(t)) - 29.24Acid%$^{0.25}$ - 279CaMg^2 + 10.63√Ol%

$$(r^2=0.79, \quad n=30) \tag{4.10}$$

where Cs-pe(t)=Cs in 1+perch in year t (Bq/kg ww); Vol=lake volume (km^3); BA=area of accumulation bottoms (%); Cs-bo(t)=Cs in bottom sediment traps in year **t-1**, thus not in the same year as for pike but in the previous year (Bq/kg dw); Acid%=percentage of acidic bedrocks in the catchment area (%); CaMg=hardness of the lake water (meq/l; mean value from the production period in year $t-1$; Ol%=percentage of open land in the catchment area (%).

The formula may be used when the parameters are within the following ranges. **Subsequently, it will also be been used in an excercise to predict the future decrease in Cs-pi.**

Cs-pe	Vol	BA	Cs-bo	Acid%	CaMg	Ol%
1500-96 000	0.0002-0.004	35.3-86.4	410-57 600	0-100	0.112-0.532	0-20

It should be observed that this is an ecometric, correlative formula designed to statistically explain variations between the lake ratio Cs-pi/Cs-pe$^{0.4}$. In this case, it leads to a consequence that would be remarkable in a dynamic context, namely that Cs-pi decreases when Cs-bo increases. In fact, it is the ratio Cs-pi/Cs-pe$^{0.4}$ that decreases when Cs-bo increases. The decrease in Cs-pi when Cs-bo increases is compensated in Eq. (4.10) by Cs-pe decreasing to a relatively greater extent.

Table 4.22 shows how data on Cs-pi can be calculated according to Eq. (4.10) for 1987 and 1988. The relationship between empirical data and model data for these 2 years is given in Fig. 4.15. As is shown by Fig. 4.15, the model gives a satisfactory description of Cs-pi for these 2 years. It is based on four "constant" environmental parameters (Acid%, Ol%, Vol and BA) and on three parameters which vary with time (Cs-pe, Cs-bo and CaMg). We now have the ecometric models required to predict the future development as regards Cs-pi. That prognosis is then based on the following conditions:

• The values as regards Cs-pe are taken from Table 4.18B, i.e., Cs-pe data have been corrected for natural radioactive decay.

• The values on Cs-bo are taken from Table 4.12, i.e., they include both natural radioactive decay and the delay factor of 1.5.

• The values placed on CaMg have not been varied. Here we assume that the hardness in any given lake in the future will be the same as the mean value for the period 1986-1988.

• The other parameters included are "constants".

From Table 4.23, which gives the final result, we can see that the ecometric modelling shows that the Cs-pi values will decrease slowly in many lakes. We also know the factors that influence the time process (from Eq. 4.10 and from the kpe and kbo equations). Figure 4.16 gives recovery curves for the lake (among our 15 lakes) that will recover slowest (lake 2201)

Table 4.22. Information on Eq. (4.10). Included parameters (Vol, BA, Ol%, Acid%, CaMg, Cs-bo and Cs-pe) and a comparison between empirical data and model data.

Lake	Vol (km^3)	BA (%)	Ol% (%)	Acid% (%)	CaMg (meq/l)	Cs-bo (Bq/kg dw)	Cs-pe (Bq/kg ww)	Cs-pi-emp (Bq/kg ww)	Cs-pi-mod (Bq/kg ww)
1987									
2110	0,0020	86,4	0	95	0,30	8400	2300	322	132
2117	0,0020	86,1	1	56	0,17	36700	16900	942	1038
2119	0,0004	73,7	0	100	0,11	19200	12700	821	883
2120	0,0003	78,1	0	96	0,25	3300	5200	260	896
2121	0,0003	81,7	9	100	0,24	37700	25700	2009	1042
2122	0,0005	85,0	0	21	0,13	29800	27000	1191	1621
2201	0,0020	35,3	0	5	0,15	576000	51800	4654	5322
2206	0,0040	80,2	1	42	0,14	68100	23500	1932	2077
2212	0,0010	78,4	0	1	0,16	241000	76200	4147	3260
2213	0,0020	81,4	3	0	0,21	137000	96000	4444	8753
2214	0,0002	72,1	1	0	0,13	115000	49510	6540	7127
2215	0,0010	78,4	7	86	0,19	95100	23800	1683	931
2216	0,0010	75,9	11	5	0,40	56200	44794	2692	3351
2217	0,0010	78,4	20	7	0,53	19100	6400	769	1145
2218	0,0010	84,6	8	97	0,30	6400	9100	818	1371
1988									
2110	0,0020	86,4	0	95	0,37	1060	1502	323	459
2117	0,0020	86,1	1	56	0,27	670	10600	2088	2930
2119	0,0004	73,7	0	100	0,11	1600	5500	1359	1874
2120	0,0003	78,1	0	96	0,37	4400	1754	791	90
2121	0,0003	81,7	9	100	0,21	410	10618	3909	3824
2122	0,0005	85,0	0	21	0,22	800	7951	2821	2772
2201	0,0020	35,3	0	5	0,24	18520	33100	7881	7343
2206	0,0040	80,2	1	42	0,12	5312	18100	4382	4011
2212	0,0010	78,4	0	1	0,26	10890	41600	6212	5244
2213	0,0020	81,4	3	0	0,21	20973	23400	9687	6703
2214	0,0002	72,1	1	0	0,20	15590	30000	7380	7403
2215	0,0010	78,4	7	86	0,25	10878	20900	2423	2335
2216	0,0010	75,9	11	5	0,40	23145	24570	3535	3442
2217	0,0010	78,4	20	7	0,52	7330	4986	•	1602
2218	0,0010	84,6	8	97	0,32	5246	2707	916	862

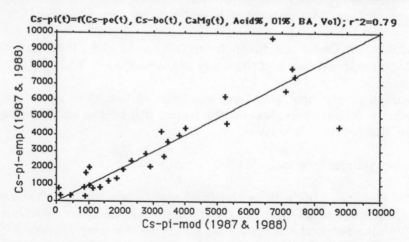

Fig. 4.15. Comparison between empirical data (y-axis) on Cs-pi for 1987 and 1988 and model data (from Eq. 4.10) on the x-axis. The model gives an r^2-value of 0.79 (n=30).

Table 4.23. Prognosis of the development as regards caesium in pike up to 2030.

Cs-pi in Bq/kg ww - with delay for Cs-bo and natural decay.								
Lake	Emp 1987	Mod 1987	Emp 1988	Mod 1988	1990	1995	2000	2010	2030
2110	322	132	323	459	877	979	451	112	7
2117	942	1038	2088	2930	1250	1080	733	225	10
2119	821	883	1359	1874	2762	2855	1184	309	21
2120	260	896	791	90	988	810	311	46	1
2121	2009	1042	3909	3824	3653	4513	2846	1090	160
2122	1191	1621	2821	2772	3231	2838	1059	197	7
2201	4654	5322	7881	7343	11918	13012	5541	2122	311
2206	1932	2077	4382	4011	4019	4331	2573	853	94
2212	4147	3260	6212	5244	6675	6389	3129	816	55
2213	4444	8753	9687	6703	7678	4380	1521	193	3
2214	6540	7127	7380	7403	9593	8700	4173	1319	132
2215	1683	931	2423	2335	2820	2385	981	174	6
2216	2692	3351	3535	3442	3256	2278	1164	179	4
2217	769	1145	•	1602	1075	977	739	322	27
2218	818	1371	916	862	729	684	524	241	28

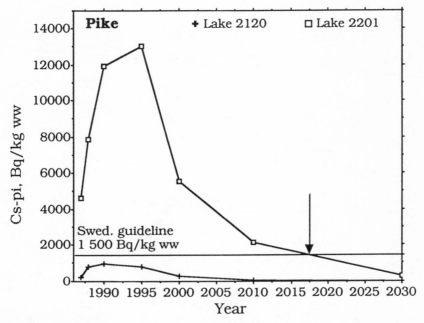

Fig. 4.16. Predicted recovery for caesium in pike in the lake which, according to the ecometric model, will recover slowest (2201) and the lake predicted to recover fastest (2120) among our 15 lakes. The other lakes fall within the range given in the figure.

and the one that will recover fastest (lake 2120). This figure shows that we will have to wait until about year 2020 until pike from lake 2201 can be eaten without the consumers having to worry about the guideline (of 1500 Bq/kg ww) given by the National Swedish Food Administration.

As already stated, there are several uncertainties in this prediction, e.g.:

• Does the delay constant 0.4 give the most relevant recovery and link between Cs-pi and Cs-pe?

• How can the models be applied to other lakes?

We intend to test these ecometric results against results from dynamic modelling (Chap. 5) and also to re-run these calculations when new data from 1989 become available. **In this context, we have aimed foremost at illustrating the possibilities (and weaknesses) of ecometric modelling.**

5. DYNAMIC MODELLING

5.1. PRESUPPOSITIONS

The aim of this chapter is to discuss a new dynamic model (for mainly caesium in pike), i.e., a model for the <u>flow</u> of caesium based on calculations using differential equations. The dynamic models used, e.g., by Carlsson (1978) or Calmet et al. (1990; see Fig. 5.1), are based on detailed information on the caesium flow at several levels in a given well-studied lake. The objective here is not to discuss more traditional dynamic models, like the one illustrated in Fig. 5.1, but a model to predict the caesium flow in lakes for which **only a minimum** of dose and sensitivity data is available.

This model is first based on the assumption that the Cs-concentration in pike (Cs-pi) can be predicted from the Cs-concentration in perch (Cs-pe), which in turn depends on the primary dose (Cs-soil) and on environmental parameters describing the catchment area and the lake. Cs-pi at a certain time (t) can be obtained from the following box model (see Fig. 5.2):

$$d[\text{Cs-pi}]/dt = k_5 * [\text{Cs-pe}] - (k_{1/2} + k_6) * [\text{Cs-pi}] \tag{5.1}$$

where Cs-pi=Cs-concentration in pike (Bq/kg ww); Cs-pe=Cs-concentration in 1+perch (Bq/kg ww); k_5=caesium uptake in pike (1/month); $k_{1/2}$=physical decay (1/month); k_6=caesium release from pike (1/month).

This means that we only use two of the boxes in Fig. 5.1 in this model, namely box 5 (top predator=pike) and box IV (small fish=perch fry). That is, it is assumed that the Cs-intake to pike from all types of prey (roach, larger perch, etc) may be estimated from only one prey, 1+perch.

The Cs-concentration in 1+perch (Cs-pe) at a certain time (t) can be written:

$$d[\text{Cs-pe}]/dt = [\text{Cs-soil}] * k_3 * k_{dev} - (k_{1/2} + k_4) * [\text{Cs-pe}] \tag{5.2}$$

where Cs-soil=fallout (Bq/m^2); k_3=the fallout constant (1/month); k_4=caesium release from perch (1/month).

kdev is important in this context. This is a **dimensionless moderator** which is defined by the relationship:

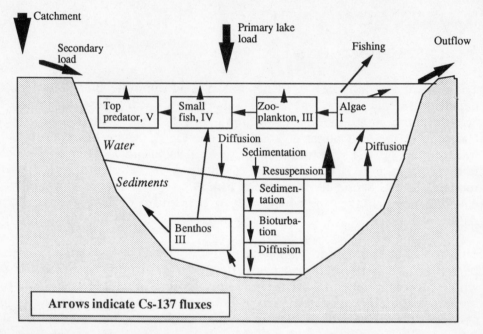

Fig. 5.1. Basic components of a general dynamic model to calculate the flux of Cs-137 to a top predator in a limnic ecosystem. The arrows correspond to rates (from Calmet et al., 1990).

$$k_{dev} = {}^n\pi\, Y_i \tag{5.3}$$

where π=the product; n=the number of environmental factors; Y_i=the environmental factor Y, where i=1, 2, ...n; here we have tested two (n=2) environmental factors, namely the theoretical water turnover time (T) and the hardness of the lake water (CaMg).

Equation (5.4) is an **algorithm** for expressing an environmental factor of this kind.

$$Y = 1 + X_{div}*([X]/X_{1/2} - 1) \tag{5.4}$$

where X_{div}=the amplitude value of the environmental factor (for, e.g., T and CaMg, see Fig. 5.3 for illustration); [X]=measured value of the environmental factor; $X_{1/2}$=normal value of the environmental factor ($\approx$median value).

For example, for the theoretical water turnover time (T), Eq. (5.4) may be written: $Y_T=1 + T_{div}*([T]/T_{1/2} - 1)$.

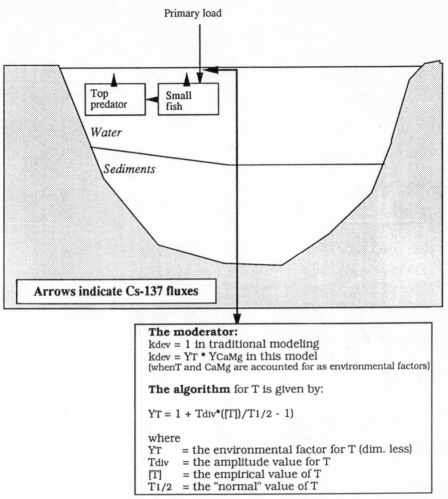

Fig. 5.2. Basic components of a dynamic model to calculate the flux of Cs-137 to pike and perch fry using a dimensionless moderator which accounts for the influence from environmental factors, such as the theoretical water turnover time (T) and the lake water hardness (CaMg).

Empirical caesium data for all 41 lakes have been used in determining the constants k_3, k_4, k_5 and k_6 in Eqs (5.1) and (5.2). The parameter k_{dev} has, in the first approach, been placed equal to one (=1) for all lakes, after which k_3 and k_4 have been calculated by minimizing the so-called **N2-norm**, which is defined by:

$$N2\text{-norm} = 1/n * \Sigma (Cs_{emp} - Cs_{mod})^2 \tag{5.5}$$

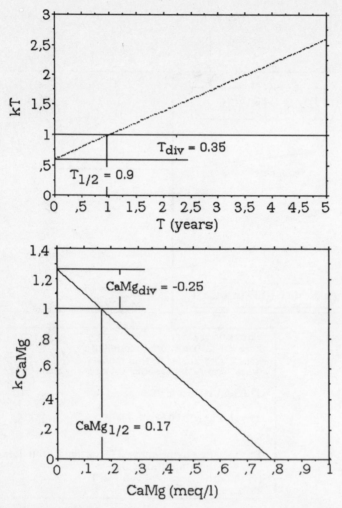

Fig. 5.3. Graphical presentation of (top) how the moderator for the water turnover time (k_T) can be defined and linked to the "median water turnover time" (=standard value=$T_{1/2}$) and the amplitude value for the water turnover time (T_{div}) and (below) how the moderator for the hardness of the water (k_{CaMg}) is defined and linked to the median value ($CaMg_{1/2}$) and the amplitude value ($CaMg_{div}$).

where n=number of observations; Cs_{emp}=empirical value placed on Cs-137 in fish (pe=perch and/or pi=pike), in Bq/kg ww; Cs_{mod}=model value placed on Cs-137 in fish in Bq/kg ww.

After the constants k_3 and k_4 have been minimized with regard to the N2-norm, they are entered into Eq. (5.1), after which k_5 and k_6 are determined. For $k_{dev}=1$ this calibration gives k_3=1.45, k_4=0.045, k_5=0.009 and k_6=0.025. The N2-norm is then 229 for perch and 4.5 for pike.

Table 4.17 shows that mainly the proportion of acidic bedrocks in the catchment area, the water turnover time, factors linked to bottom dynamic conditions and the water chemical cluster parameters appear to be important sensitivity parameters for the recovery process. The question is how to use these parameters in our calculations? In order to investigate this, the model has first been calibrated with regard to each of these factors individually. For each variable (Y) different values have been placed on $X_{1/2}$ and X_{div} in Eq. (5.2) - Eq. (5.5). The best pair of values with regard to the N^2-norm is shown in Table 5.1. We can see that the N^2-norm for perch decreases in the size range from 229 to 150 if the environmental factors are allowed to influence the model. If T and CaMg are combined according to Eq. (5.3) the N^2-norm$_{pe}$ and the N^2-norm$_{pi}$ will be 142 and 2.9, respectively. This combination has been used in the calculations below.

Table 5.1. X1/2, Xdiv and N^2-norm for perch (pe) and pike (pi) for different sensitivity parameters (Acid%=the percentage of acid bedrocks in the catchment area, BA=the area of accumulation bottoms in the lake, T=the theoretical water turnover time, CaMg=hardness of lake water, and cond=conductivity of lake water).

Parameter	$X_{1/2}$	X_{div}	N^2norm$_{pe}$	N^2norm$_{pi}$
No (k_{dev}=1)	1	0	229	4.5
Acid %	30	-0.05	227	4.5
BA	75	0.8	209	4.1
T	0.9	0.35	150	3.0
CaMg	0.17	-0.25	186	3.2
Cond	2.4	-0.35	183	3.4

Equation (5.4) describes how one can account for how the water turnover time (T) influences the recovery in Cs-pe. The relationship gives a normal value of 1 for lakes with T=0.9 year. The more T diverges from 0.9 year the more the k_T-value diverges from 1 (see Fig. 5.2). A water turnover time of 0.9 year may be regarded as a normal value around which the model swings. The amplitude is given by the T_{div}-value, which has been calibrated to 0.35 for perch. In the same way as for T, corresponding values are determined for CaMg (see Fig. 5.2). For CaMg, 0.17 (meq/l) is used as normal value (CaMg$_{1/2}$); the amplitude value (CaMg$_{div}$) is given by -0.25.

5.2. SIMULATION IN INDIVIDUAL LAKES

How does this model function for a given lake? In this section, we discuss this using data from lake 2201 (Selasjön). We first simulate the future development for perch and pike assuming that the conditions prevailing before the remedial measures will continue. The results are shown in Fig. 5.4. Here the following data have been used: Cs-soil=70 kBq/m^2, T=0.33 years, CaMg=0.14 meq/l. This gives a k_{dev}-value of 0.74. Figure 5.4 shows the recovery curves for perch and pike together with data for Cs-pe and Cs-pi for a few selected years after the Chernobyl event.

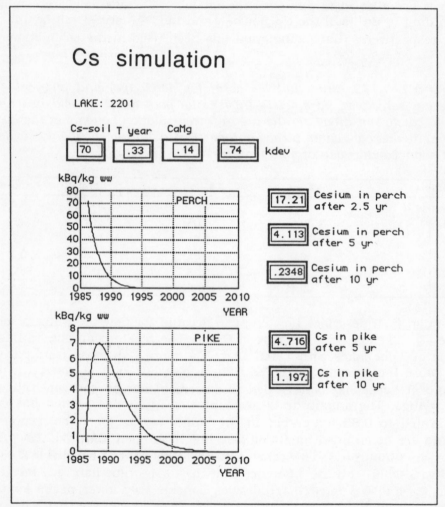

Fig. 5.4. Simulation of how the Cs-concentrations in perch and pike can be predicted by the dynamic model in a selected lake (2201), where the starting parameters are Cs-soil=70 kBq/m^2, water turnover time T=0.33 years, and water hardness 0.14 meq/l. The Cs-values 2.5, 5 and 10 years after Chernobyl (i.e., 1986) are given in boxes.

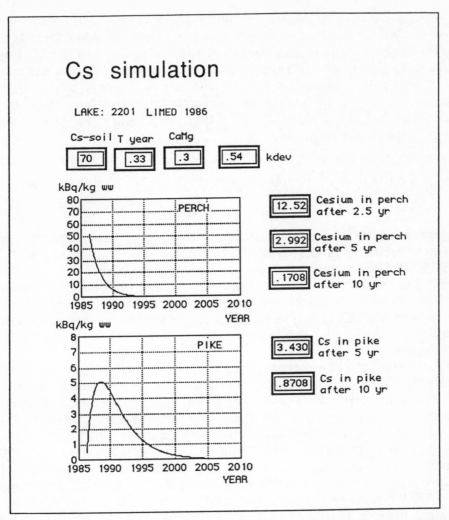

Fig. 5.5. Simulation of an effect of a remedial measure. The same conditions apply as in Fig. 5.4 except that the model simulates a probable effect of a liming which increases the hardness of the water from 0.14 meq/l to 0.3 meq/l. The recovery will then be faster for both 1+perch and pike.

Figure 5.5 gives a prognosis of what might happen had lake 2201 been limed in 1986. In this case, the same model conditions have been used as in Fig. 5.4 but with the difference that the CaMg-value has been increased from 0.14 to 0.3 (meq/l). This gives considerably lower peaks for both perch and pike and a faster recovery. For pike we can see that the guideline of 1.5 kBq/kg ww will be reached already in 1993/94.

The dynamic model appears to give a faster recovery than the ecometric model. Which model is most reliable? Today it is pointless in speculating in greater detail on this matter, although it seems highly probable the present dynamic models yields too fast a recovery for pike. **The dynamic model can be extended and improved and, naturally, has the best opportunities to give a scientifically relevant description of this complicated time process.** In future work, we will also test alternatives to the N^2-norm, e.g., the N^1-norm, the $N^{1/2}$-norm and alternative expressions for the N^2-norm, like: N^2-norm $=1/n*\Sigma[(Cs_{emp}-Cs_{mod})/Cs_{mod}]^2$. But that is outside the framework of this publication.

5.3. COMPARISON BETWEEN ECOMETRIC AND DYNAMIC MODELLING

Figure 5.6 gives direct comparisons between empirical data and model-predicted data from the dynamic model for 1987 (Fig. 5.6A) and 1988 (Fig. 5.6B). The degrees of explanation are high for the 2 years — 79% for 1987 data and 78% for 1988 data. The model predictions coincide satisfactorily with the ideal y=x line in 1987 but not in 1988, when the dynamic model starts to give values that are too low. Is this also so for the perch?

We have tried to answer that question in Fig. 5.7. This figure gives the relationship between empirical data and model-calculated data according to the dynamic model for Cs-pe86. The r^2-value is 0.69, which is a high value for the fallout year, when conditions were exceptionally variable. The r^2-value was higher in 1987 when 79% of the variation in the empirical values can be explained by the dynamic model (Fig. 5.7B). For 1988, the dynamic model describes the conditions fairly well in all lakes but one (2203; Fig. 5.7C). If this lake is eliminated, the r^2-value will rise from 0.68 to 0.78. We can also see from the diagrammes in Fig. 5.7 that, in contrast to what appears to apply to pike, there is no trend indicating that the dynamic model would predict too rapid recovery for perch. That the model predictions are probably better for perch than for pike are not unexpected since **the constants included have been calibrated for 3 years of empirical data for perch and only for 2 years of data for pike**. This is a significant difference in this connection, and it is probable that a much more reliable dynamic model may be obtained when data from more years become available.

Figure 5.8 compares results from the dynamic and the ecometric model directly for Cs-pi. From Fig. 5.8A, we can see that both the dynamic and ecometric models give data for 1987 that coincide very well with the y=x line. One lake (2213) clearly diverges. **A divergence of this kind for a single lake is not remarkable; it may depend on deficiencies in the empirical material just as well as in the models.**

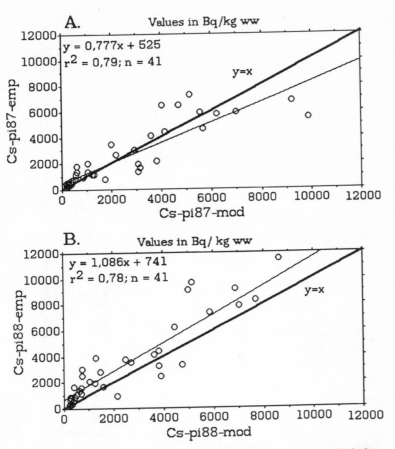

Fig. 5.6. A. Relationship between empirical data and model data from the dynamic model for caesium in pike from 1987.
B. Relationship between empirical data and model data from the dynamic model for caesium in pike from 1988. The figure gives regression lines, their equations, degrees of explanation (r^2) and y=x lines.

A corresponding comparison for 1+perch is given in Fig. 5.9. Three points can be identified as diverging. This concerns data from lake 2203 for the dynamic model and data for lake 2207 for both the dynamic and the eco-metric model. If we omit these diverging data, the adaptation to the ideal line is satisfactory. Figure 5.9B shows corresponding comparisons for 1986 for the 15 matrix lakes (since we have only processed ecometric model data for 1987 and 1988 for these 15 lakes). We can see that the ecometric model generally predicts higher Cs-concentrations in 1+perch than the dynamic model and that the differences appear to increase from year to year. This is one further indication that the ecometric model pre-

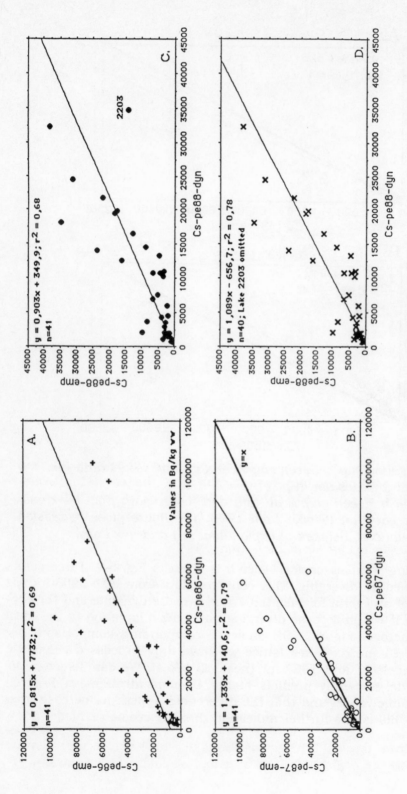

Fig. 5.7. A. Relationship between empirical data and model data from the dynamic model for caesium in
1+perch from 1986 for all 41 lakes.
B. The same for the 1987 material; here the y=x line is also given.
C. The same for the 1988 material (n=41).
D. The same for the 1988 material when lake 2203 is omitted.
The figure gives regression lines, their equations, the number of lakes (n) and r²-values.

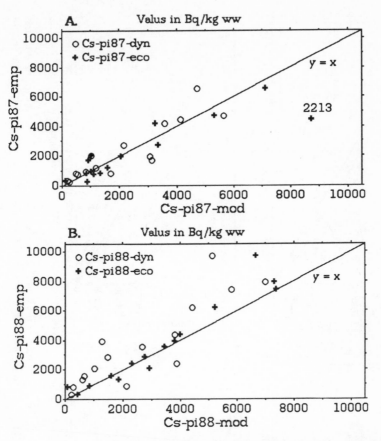

Fig. 5.8. A. Relationship between empirical data (emp) and model data (mod) from both the dynamic model (dyn) and the ecometric model (eco) for caesium in pike from 1987 (Cs-pi87) for data from the 15 lakes. In the figure, the outlier lake 2213 has been marked, as well as the y=x line.
B. The same as above but for the data from 1988.

dicts a slower recovery than the dynamic model. The models predict Cs-concentrations in pike and 1+perch satisfactorily for the years for which they have been calibrated.

From Table 5.2, giving the final results for all 41 lakes for the dynamic modelling for Cs-pi, and from Table 5.3 giving corresponding data for 1+perch, we can see that the Cs-values will decrease slowly in many lakes. We also know the factors that influence the time process, mainly fallout, lake turnover time and hardness of the lake water.

Figure 5.10 gives recovery curves for the lake which will recover slowest (lake 2203) and the one that will recover fastest (lake 2110) according to

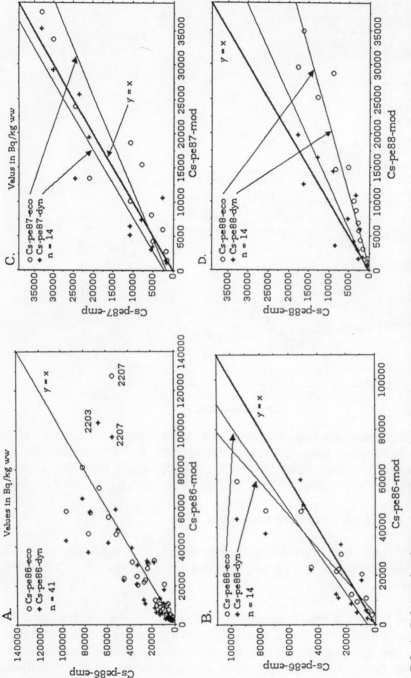

Fig. 5.9. A. Relationship between empirical data (emp) and model data (mod) from both the dynamic model (dyn) and the ecometric model (eco) for caesium in 1+perch from 1987 (Cs-pe87) for the 41 lakes. In the figure, the outliers have been marked as well as the y=x line.
B. The same as above but for the data from1986 from the 15 lakes. The figure gives the regression lines between empirical data and dynamic data and between empirical data and ecometric data, as well as the y=x line.
C. The same as for B but for the data from1987 from the 15 lakes.
D. The same as for B but for the data from 1988 from the 15 lakes.

Table 5.2. Result of computer simulations according to the dynamic mo-del for how the Cs-concentrations in pike might develop up to year 2030. The values (both empirical data=emp, and model data=mod) are given in Bq/kg wet weight.

Dynamic model, simulations for pike. Values in Bq/kg ww.

Lake	1987 emp.	1987 mod.	1988 emp.	1988 mod.	1990	1995	2000	2010	2030
2101	1158	1289	1679	1596	1318	367	79	3	0
2102	195	209	304	259	214	60	13	1	0
2103	1358	1028	1982	1273	1051	293	63	3	0
2104	1132	562	1178	695	574	160	34	1	0
2105	531	323	728	400	330	92	20	1	0
2106	297	264	329	328	270	75	16	1	0
2107	3483	2003	3746	2481	2049	570	122	5	0
2108	743	424	1088	526	434	121	26	1	0
2109	417	285	612	353	292	81	17	1	0
2110	322	187	323	231	191	53	11	0,5	0
2111	258	197	351	244	201	56	12	0,5	0
2112	623	323	1663	400	330	92	20	1	0
2113	1784	607	3014	752	620	173	37	2	0
2114	400	369	837	457	378	105	23	1	0
2115	285	216	616	267	221	61	13	1	0
2116	537	203	844	252	208	58	12	0,5	0
2117	942	837	2088	1036	856	238	51	2	0
2118	1346	608	2497	753	622	173	37	2	0
2119	821	494	1359	612	505	141	30	1	0
2120	260	243	791	300	248	69	15	1	0
2121	2009	1040	3909	1288	1063	296	64	3	0
2122	1191	1205	2821	1493	1232	343	74	3	0
2201	4654	5667	7881	7019	5795	1613	346	14	0
2202	5797	6212	8344	7695	6353	1769	380	15	0
2203	5476	9898	8863	12260	10121	2818	605	24	0
2204	3016	2932	4116	3631	2998	835	179	7	0
2205	1355	3065	3217	3797	3135	873	187	8	0
2206	1932	3066	4382	3797	3135	873	187	8	0
2207	6813	9214	13756	11413	9422	2623	563	23	0
2208	5932	5555	9164	6880	5680	1581	340	14	0
2209	2194	3804	3335	4712	3890	1083	233	9	0
2210	7310	5150	14721	6380	5267	1466	315	13	0
2211	5919	6974	11473	8638	7131	1985	426	17	0
2212	4147	3586	6212	4442	3667	1021	219	9	0
2213	4444	4145	9687	5135	4239	1180	253	10	0
2214	6540	4707	7380	5830	4813	1340	288	12	0
2215	1683	3140	2423	3890	3211	894	192	8	0
2216	2692	2163	3535	2679	2212	616	132	5	0
2217	769	560	1602	694	573	159	34	1	0
2218	818	1722	916	2133	1761	490	105	4	0
2219	6483	4026	9102	4986	4116	1146	246	10	0

Table 5.3. Result of computer simulations according to the dynamic mo-del for how Cs-concentrations in 1+perch might develop up to year 2030. The values (both empirical data=emp, and model data=mod) are given in Bq/kg wet weight.

Dynamic model, simulations for perch. Values in Bq/kg ww.

Lake	1986 emp.	1986 mod.	1987 emp.	1987 mod.	1988 emp.	1988 mod.	1990	1995	2000	2010	2030
2101	12900	13540	7295	8010	1961	4518	1659	95	5	0,02	0
2102	2800	2200	514	1302	1083	734	269	15	1	0	0
2103	9900	10799	3289	6389	2706	3603	1323	76	4	0,01	0
2104	11300	5900	8654	3490	2069	1969	723	41	2	0,01	0
2105	7200	3393	5593	2007	812	1132	416	24	1	0	0
2106	3390	2779	1802	1644	578	927	340	19	1	0	0
2107	33400	21050	19355	12453	6574	7024	2578	147	8	0,03	0
2108	5600	4459	4232	2638	1942	1488	546	31	2	0,01	0
2109	8200	2997	2500	1773	415	1000	367	21	1	0	0
2110	2300	1961	1502	1160	451	654	240	14	1	0	0
2111	1730	2066	1236	1222	1077	689	253	14	1	0	0
2112	7160	3396	5100	2009	3412	1133	416	24	1	0	0
2113	13300	6375	7544	3772	2936	2127	781	45	3	0,01	0
2114	2200	3880	1688	2295	1160	1295	475	27	2	0,01	0
2115	4600	2269	1826	1342	739	757	278	16	1	0	0
2116	4200	2138	1637	1265	359	713	262	15	1	0	0
2117	16900	8792	10600	5202	2857	2934	1077	61	4	0,01	0
2118	15800	6386	12480	3778	9916	2131	782	45	3	0,01	0
2119	12700	5191	5500	3071	2674	1732	636	36	2	0,01	0
2120	5200	2548	1754	1508	586	850	312	18	1	0	0
2121	25700	10927	10618	6464	8393	3646	1338	76	4	0,01	0
2122	27000	12661	7951	7490	3852	4225	1551	89	5	0,02	0
2201	51800	59541	33100	35224	17588	19868	7293	416	24	0,08	0
2202	81200	65274	83300	38616	22139	21781	7996	456	26	0,08	0
2203	66616	103998	42800	61525	14274	34702	12739	727	42	0,1	0
2204	32768	30805	17700	18224	3426	10279	3773	215	12	0,04	0
2205	17577	32208	14600	19054	4400	10747	3945	225	13	0,04	0
2206	23500	32211	18100	19056	6521	10748	3946	225	13	0,04	0
2207	54334	96813	97900	57274	38206	32305	11859	677	39	0,1	0
2208	74299	58365	55000	34529	18470	19475	7149	408	23	0,08	0
2209	37400	39969	23000	23646	4611	13337	4896	279	16	0,05	0
2210	53308	54116	57700	32015	34731	18057	6629	378	22	0,07	0
2211	70419	73272	77724	43347	31302	24449	8975	512	29	0,1	0
2212	76200	37682	41600	22292	16335	12574	4616	263	15	0,05	0
2213	96000	43557	23400	25768	8919	14534	5335	305	17	0,06	0
2214	49510	49455	30000	29257	12839	16502	6058	346	20	0,06	0
2215	23800	32996	20900	19520	3250	11010	4042	231	13	0,04	0
2216	44794	22726	24570	13445	5259	7583	2784	159	9	0,03	0
2217	6400	5886	4986	3482	1487	1964	721	41	2	0,01	0
2218	9100	18090	2707	10702	2196	6036	2216	126	7	0,02	0
2219	58030	42297	34100	25023	23856	14114	5181	296	17	0,06	0

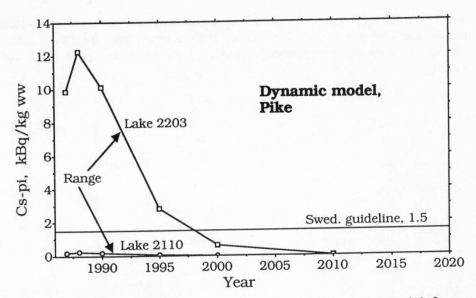

Fig. 5.10. Predicted recovery according to the dynamic model for caesium in pike in the lake that, according to the model, would recover slowest (2203) and the one which would recover fastest (2110) among our 41 lakes. The other lakes are grouped within the range given in the figure.

the dynamic modelling. From this figure, we can see that during some time between 1995 and 2000 the average pike in lake 2203 will have Cs-concentrations below the guidelines issued by the National Swedish Food Administration. Figure 5.11 gives a more direct comparison between the predictions for a lake that recovers slowly (2201), a lake that recovers "moderately rapidly" (2206) and a lake that recovers rapidly (2118). From these three recovery diagrammes, we may note:

• that the agreement between predicted data according to the two models is very good for the years for which the models are calibrated, and

• that the ecometric model generally gives a slower recovery for pike.

5.4. ECOLOGICAL HALF-LIFES

From the recovery models (see previous tables and diagrammes), we may estimate preliminary values for the **ecological half-lifes** for caesium in pike and 1+perch after the Chernobyl event. The dynamic model yields a half-life of about 7 yr (SD<1 yr) for pike and 1.7 yr for 1+perch. The ecometric models give a half-life of about 13 yr for pike (SD=2.2 yr) and about 2.1 yr for 1+perch (see Table 5.3 for perch). These ecological half-lifes have been determined relative **to the year when the maximal Cs-concentration** appear in the two species of fish for each individual lake (see Fig. 5.12). The max.- values are normally attained much later for pike than for 1+perch:

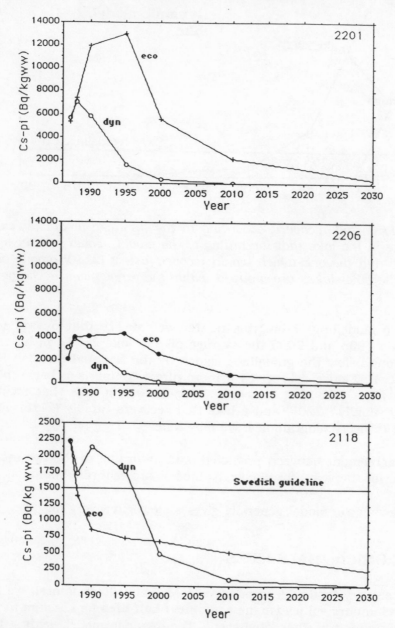

Fig 5.11. Recovery curves according to the ecometric model (eco) and the dynamic model (dyn) for the period from the Chernobyl event in 1986 until year 2030 for a lake with a slow recovery (2201 — upper), a lake with a moderately fast recovery (2206 — centre) and a lake with a rapid recovery (2118 — lower). The guideline established by the National Swedish Food Administration (1500 Bq/kg wet weight) is also given in the lower figure.

Table 5.4. An example how ecological half-lifes for caesium in 1+perch after Chernobyl have been determined for 15 Swedish lakes from ecometric modelling.

Ecological half-life (t1/2) for Cs in 1+perch according to ecometric model.										
Lake	1986	1987	1988	1990	1995	2000	2010	2030	Max. year	t1/2 year	t1/2
2110	3939	1446	1021	521	82	13	0	0	1986	1987	1
2117	12201	10128	6899	3318	434	57	1	0	1986	1988	2
2119	9157	8114	5799	3021	502	83	2	0	1986	1988	2
2120	5395	2742	1697	695	56	5	0	0	1986	1987	1
2121	22753	18662	14680	8975	2410	647	47	0	1986	1990	4
2122	21634	15423	10133	4590	501	55	1	0	1986	1988	2
2201	46439	37615	29589	18091	4857	1304	94	0	1986	1990	4
2212	46904	48685	34791	18126	3011	500	14	0	1987	1990	4
2213	58721	48248	28799	11116	751	51	0	0	1986	1988	2
2214	48280	33723	25284	14270	3014	636	28	0	1986	1988	2
2215	28994	13379	8686	3856	396	41	0	0	1986	1987	1
2216	23480	24000	15030	6284	540	46	0	0	1987	1988	2
2217	10611	4222	3090	1676	314	59	2	0	1986	1987	1
2218	20667	6003	4394	2383	446	84	3	0	1986	1987	1
							Mean:		1986,14	1988,07	2,07
							SD:		0,36	1,14	1,14
							Min.		1986	1987	1
							Max.		1987	1990	4

• for pike, 1991 (SD=3.2 yr), i.e., on average 5 years after the fallout, according to the ecometric model and 1988 (SD<1 yr) according to the dynamic model;

• for 1+perch, fall of 1986, i.e., about half a year and an entire summer after the fallout, according to both the ecometric and the dynamic models.

It should be stressed that the ecological half-life, in contrast to the physical half-life (30.2 years for Cs-137), is not a constant. The ecological half-life depends on the species of fish; and for the same species, it is different in different lakes, depending on, e.g., lake water turnover time, bottom dynamic conditions and water chemistry.

> The intention with this section has not been to give "final" results on the recovery process in fish. The ambition has been to present the general design of the models and to discuss some of their inherent deficiencies and advantages.

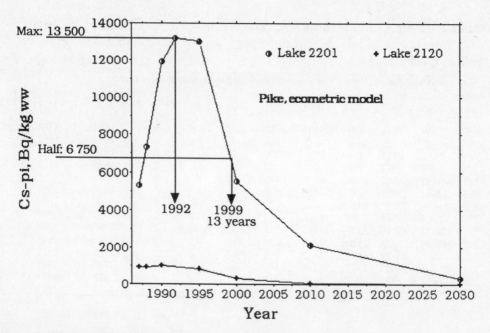

Fig. 5.12. Illustration of the concept ecological half-life. The upper curve shows the situation in lake 2201 (as determined by the ecometric model). The max-value for Cs-pi is attained in 1992 (i.e., 6 years after the Chernobyl accident) and the ecological half-life for pike in this lake is 13 years. The lower curve illustrates that the conditions are quite different for pike in lake 2120.

6. THE GEOGRAPHICAL PERSPECTIVE

As an important part in a general analysis of environmental consequences, one should not only study relationships between dose, sensitivity and effect parameters, but also attempt to make a problem dimensioning. A central aspect of problem dimensioning is to try to establish the size of the impact areas of substances causing environmental contamination; large impact areas generally imply large potential ecological risks, and vice versa (see Håkanson, 1984a).

Based on empirical data from 644 Swedish lakes, Andersson et al. (1990) have presented a map showing the distribution pattern of caesium in Swedish lake fish after Chernobyl. That report will not be discussed here. Our objective in this chapter is, instead, to use the results from our 41 lakes from a limited area of Sweden and extend them into a larger geographical context. Thus, it will be exemplified for 1+perch and pike how results from ecometric analysis can be used to get a geographical perspective on the caesium situation in Swedish lake fish.

It must be stressed that the use of ecometric models on **individual** lakes outside the range of the models may lead to incorrect results. However, it is quite a different matter to attempt to obtain large, **regional**, geographical patterns for many lakes. Then it need not be of decisive importance to make a correct prediction of the conditions in each lake. In addition, there are few, if any, economically and practically feasible alternatives to ecometric models for predicting major geographical distribution patterns where the fundamental unit is a lake.

6.1. CAESIUM IN 1+PERCH

The following ecometric formula (from Håkanson et al., 1988) gives, in quantitative terms, the relationship between fallout, lake sensitivity and effect (i.e., Cs-137 in 1+perch):

$$\log(\text{Cs-pe86/87}) = 0.975*\log(\text{Cs-soil}) - 0.185*\text{cond}* + 0.545 \qquad (6.1)$$
$$(r^2=0.80, \ n=41)$$

where Cs-pe86/87=caesium in 1+perch in Bq/kg (dry weight and not wet weight); lake typical mean for 1986 and 1987; Cs-soil=fallout in Bq/m^2; cond*=mean conductivity in mS/m (millisiemens per metre) before remedial measures.

The formula, which gives a degree of explanation of 80% of the variability in the effect parameter between the 41 lakes, states that lakes with low conductivity have great sensitivity, i.e., that in lakes with low cond-values the 1+perch in 1986/87 had higher Cs-concentration, at comparable primary doses, than in lakes with high cond-values.

Monitor (1986) gives a map showing how the conductivity varies in Swedish lakes. Somewhere in the interval 3-7, say at 5 mS/m, we can place a practical rule-of-thumb limit which states that lakes with a conductivity below this limit are particularly sensitive as regards radioactive caesium. The isolines from the Monitor map for conductivity equal to 5 (mS/m) have been transferred to the map in Fig. 6.1B.

The fallout map (Fig. 2.1) is used in this context to obtain a Swedish perspective as regards the Cs-dose. If we start from a sensitivity of 5 mS/m and the guideline value of 1500 Bq/kg ww or 6000 Bq/kg dw, we can calculate from Eq. (6.1) that these values correspond to a primary dose of 14 390 Bq/m^2. The isoline for 15 000 Bq/m^2 from Fig. 2.1 has then been transferred to Fig. 6.1A. If the two maps for primary dose (Fig. 6.1A) and sensitivity (Fig. 6.1B) are pooled, we get a map (Fig. 6.1C) showing the areas in Sweden in 1986/87 that had lakes that are likely to have 1+perch with Cs-concentrations higher than 1500 Bq/kg ww. It should be noted that this map only applies to **one conductivity** value and **one value** placed **on Cs-soil**. The Cs-concentration in 1+perch may, naturally, be above the guideline value for other combinations of Cs-soil and conductivity.

A map such as Fig. 6.1C must not be interpreted to imply that all lakes within the shaded area must have fish with values above the guideline value and that all lakes outside the shaded area should "go free". Instead, the map is based on the given conditions and therefore only shows the most probable situation. The areas in Sweden in which the lakes do actually have fish with average Cs-concentrations in excess of the guideline values are reported by Andersson et al. (1989).

We have also made an overall calculation of how many lakes there are within the contaminated area. The calculations are based on the number of lakes in different counties and the lake density in these counties (see Andersson et al., 1987). The calculations indicate that there are a total of about 20 000 lakes within the contaminated area in Sweden (there are totally about 83 000 lakes in Sweden). If we assume that not all but more than half, let us say 75%, of the lakes in this area had 1+perch with values in excess of the Swedish guideline value, then it would be about 15 000 lakes. According to the calculations made by Andersson et al. (1989), based on empirical data from 644 lakes, Sweden had about 14 000 lakes in 1987 with Cs-concentrations in 1-hg perch in excess of 1.5 kBq/kg ww.

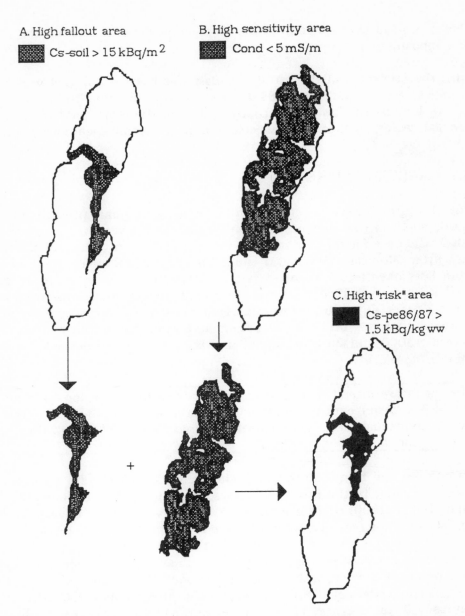

A. High fallout area

Cs-soil > 15 kBq/m^2

B. High sensitivity area

Cond < 5 mS/m

C. High "risk" area

Cs-pe86/87 > 1.5 kBq/kg ww

Fig. 6.1. A. The area in Sweden which received a larger primary dose (=fallout) than 15 000 Bq/m^2.
B. Areas with lakes with lower conductivity than 5 mS/m.
C. Areas in Sweden which, in 1986/87, had lakes where the concentration of Cs-137 in 1+perch was, on average, above the guideline level of 1500 Bq/kg wet weight as a result of the fallout being greater than 15 000 Bq/m^2 and the conductivity being lower than 5 mS/m.

This is a good agreement in terms of order of magnitude from the given presuppositions.

With the criteria available to distinguish small ecological problems from large (see e.g., Håkanson, 1984a), there is no doubt that the contamination of lake fish following Chernobyl is one of the major problems facing Swedish water management (especially from a psychological viewpoint).

6.2. CAESIUM IN PIKE

The intention here is to provide a geographical picture of the development in the Cs-concentration in pike for the years 1990, 2000 and 2010. The base for these maps is made up of empirical values of conductivity (from Monitor, 1986) and the primary dose/fallout (from Fig. 2.1) from 300 lakes (see Håkanson et al., 1990a), together with the dynamic model (Chap. 5), which gives the calculated Cs-concentrations in pike for these 3 years. Figure 6.2 shows the results of a simulation where the water turnover time has been held constant at 1 year. The primary dose varies between 2000 and 70 000 Bq/m^2 and the conductivity between 1.5 and 50 mS/m.

The maps are intended to give the mean concentrations for different regions. The mean concentration in individual lakes or the Cs-concentration in individual pike, may naturally diverge considerably from the values given in the figure.

As can be seen from Fig. 6.2, it is the areas that received a large fallout that have, and in the future will have, the highest Cs-concentrations in pike. Today (1990) there are about 7000 lakes with pike having Cs-concentrations in excess of the Swedish guideline value of 1500 Bq/kg ww.

It must be emphasized that the given dynamic model, as mentioned earlier, probably predicts a decrease in Cs-concentration in pike that is too fast. These model calculations indicate, e.g., that in year 2000 the Cs-pi-values will, on average, be in the range 100-300 Bq/kg ww within the most exposed areas in Sweden. The maximum Cs-concentrations in pike in year 2000 should be below 1500 Bq/kg ww. In the year 2010, all lakes should, in principle, have Cs-pi-values lower than 20 Bq/kg ww. It will be interesting to follow the future development. There should be good opportunities to make considerable improvements of the dynamic model when further data become available.

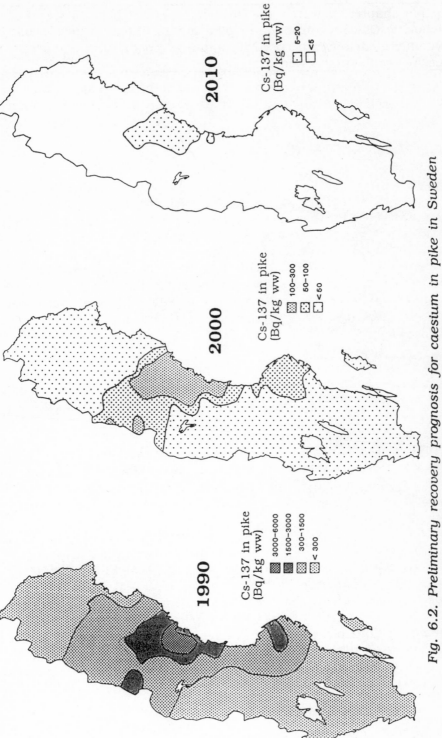

Prognosis, dynamic model

1990

Cs-137 in pike
(Bq/kg ww)

3000–6000
1500–3000
300–1500
< 300

2000

Cs-137 in pike
(Bq/kg ww)

100–300
50–100
< 50

2010

Cs-137 in pike
(Bq/kg ww)

5–20
<5

*Fig. 6.2. Preliminary recovery prognosis for caesium in pike in Sweden
according to the dynamic model for caesium in pike.*

In this chapter, however, the intention has not been to give a final prediction of Cs-concentrations in pike but to illustrate how ecometric and dynamic models can be used to obtain a geographical perspective on the caesium problem in Sweden.

7. REFERENCES

SNV=Swedish Environmental Protection Agency.

Ambio, 1990. Special Issue: Marine Eutrophication. - Vol. 19, pp. 101-176.

Andersson, T., Nilsson, Å., Håkanson, L. and Brydsten, L., 1987. Mercury in Swedish lakes (in Swedish with English summary). - SNV Rapport 3192, 92 p.

Andersson, T., Forsgren, G., Håkanson, L., Malmgren, L. and Nilsson, Å., 1990. Radioactive caesium in fish in Swedish lakes after Chernobyl (in Swedish with English summary) . - SSI Rapport 90-04, 41 p.

Bengtsson, Å., Viklund, T., Häggbom, M., Andersson, T. and Håkanson, L., 1987. Lakes and drainage areas in Västernorrland county. Project Liming-mercury-caesium (in Swedish with English summary). - SNV Rapport 3402, 81 p.

Bengtsson, Å., Göthe, L., Andersson, T. and Håkanson, L., 1988. The situation before remedial measures in in Västernorrland county. Project Liming-mercury-caesium (in Swedish with English summary). - SNV Rapport 3564, 70 p.

Black, V.S., 1957. Excretion and osmoregulation. In: The Physiology of Fishes, Brown, M.E. (ed.). - Academic Press, New York, Vol. 1, pp. 163-205.

Broberg, A. and Andersson, E., 1989. Turnover of caesium in limnic system (in Swedish). - Limnologiska inst., Uppsala univ., mimeo, 30 p.

Cairns, Jr., J. & Pratt, J.R. (1987). Ecotoxicological effect indices: A rapidly evolving system. - Wat. Sci. Tech., 19:1-12.

Carlsson, S., 1978. A model for the turnover of Cs-137 and potassium in pike (Esox Lucius). - Health Phys., 35:549-554.

Calmet, D., Håkanson, L., Bergström, U., 1990. The IAEA Coordinated Research Programme on Validation of Models for the Transfer of Radionuclides in the Aquatic Environment. - Manuscript, IAEA, Vienna.

Fleishman, D.G., 1963. Accumulatation of artificial radionuclides in freshwater fish. In: Radioecology, Klechkovskii, V.M., Polikarpov, G.G. and Aleksakhin, R.M. (editors). - John Wiley, New York, pp. 347-370.

Gustafsson, L., Lanhammar, H. and Sandblad, B., 1982. Systems and models (in Swedish). - Studentlitteratur, Lund, 275 p.

Håkanson, L., 1980. The quantitative impact of pH, bioproduction and Hg-contamination on the Hg-content in fish (pike). - Env. Poll., 1:285-304.

Håkanson, L., 1981. A Manual of Lake Morphometry. - Springer-Verlag, Heidelberg, 78 p.

Håkanson, L., 1984a. Aquatic contamination and ecological risk. An attempt to a conceptual framework. - Water Research, 18:1107-1118.

Håkanson, L., 1984b. Metals in fish and sediments from the River Kolbäcksån water system, Sweden. - Arch. Hydrobiol., 101:373-400.

Håkanson, L., 1990. An operative system for environmental consequence analysis for aquatic ecosystems. - In: Baudo, R., Giesy, J.P. and Muntau, H. (editors), Sediments: Chemistry and Toxicity of In-Place Pollutants, Lewis Publishers, Michigan, pp. 365-390.

Håkanson, L. and Jansson, M., 1983. Principles of Lake Sedimentology. - Springer-Verlag, Heidelberg, 316 p.

Håkanson, L., Andersson, T., Neumann, G., Nilsson, Å. and Notter, M., 1988. Caesium in perch in lakes from northern Sweden after Chernobyl (in Swedish with English summary). - SNV Rapport 3497, 136 p.

Håkanson, L., Floderus, S. and Wallin, M., 1989. Sediment trap assemblages - a methodological description. - Hydrobiologia, 176/177:481-490.

Håkanson, L., Kvarnäs, H., Andersson, T., Neumann, G. and Notter, M., 1990a. Caesium in pike after Chernobyl - dynamical and ecometric modelling (in Swedish). - SSI-rapport 90-09, 145 p.

Håkanson, l., Andersson, P., Andersson, T., Bengtsson, Å., Grahn, P., Johansson, J.-Å., Jönsson, C.-P., Kvarnäs, H., Lindgren G. and Nilsson, Å., 1990b. Measures against high mercury content in lake fish (in Swedish with English summary). - SNV Rapport 3818, 189 p.

Håkanson, L., Borg, H. and Uhrberg, R., 1990c. Reliability of analyses of Hg, Fe, Ca, K, P, pH, alkalinity, conductivity, hardness and colour from lakes. - Int. Rev. ges. Hydrobiol., 75:79-94.

Håkanson, L., Andersson, T. and Nilsson, Å., 1990d. A new method of quantitatively describing drainage areas. - Env. Geol. & Water Sci., 15:61-69.

Johansson, J.-Å., Lindgren, G., Andersson, I., Nilsson, Å. and Håkanson, L., 1987. Lakes and drainage areas in Gävleborg county. Project Liming-mercury-caesium (in Swedish with English summary). - SNV Rapport 3383, 90 p.

Johansson, J.-Å., Lindgren, G., Andersson, I., Nilsson, Å. and Håkanson, L., 1988. The situation before remedial measures in Gävleborg county. Project Liming-mercury-caesium (in Swedish with English summary). - SNV Rapport 3553, 80 p.

Knoechel, R. and Campbell, C.E., 1988. Physical, chemical, watershed and plankton characteristics of lakes on the Avalon Peninsula, Newfoundland, Canada: a multivariate analysis of interrelationships. - Verh. Internat. Verein. Limnol., 23:282-296.

Landner, L., (ed.), 1989. Chemicals in the aquatic environment. Advanced hazard assessment. - Springer-Verlag, Heidelberg, 350 p.

Lindqvist, O., Johansson, K., Aastrup, M., Andersson, A., Bringmark, L., Hovsenius, G., Håkanson, L., Iverfeldt, Å., Meili, M. and Timm, B., 1991. Mercury in the Swedish Environment - Recent Research on Causes, Consequences and Corrective Methods. - Textbook manuscript. Accepted for publication in Water, Air and Soil Pollution.

Mackey, D. & Paterson, S. (1982). Fugacity revisited. - Environ. Sci. Technol., 16:654-660.

Monitor, 1986. Acid and acidified waters (in Swedish). - Naturvårdsverket (SNV) Informerar, 180 p.

Neter, J., Wasserman, W. and Whitmore, G.A., 1988. Applied statistics, 3rd. Ed. - Allyn and Bacon, Boston, 1006 p.

Nilsson, Å., Andersson, T., Håkanson, L. and Andersson, A., 1989. Mercury in lake fish - linkages to mercury and selenium i the mor layer and to historical emissions (in Swedish with English summary). - SNV Rapport 3593, 117 p.

Nilsson, Å. and Håkanson, L., 1990. Relationship between drainage area characteristics and lake water characteristics. - Accepted for publication in Env. Geol. & Water Sci.

OECD, 1982. Eutrophication of waters. Monitoring, assessment and control. - OECD, Paris, 154 p.

O'Neill, R.V., Gardner, R.H., Barnthouse, L.W., Suter, G.W., Hildebrand, S.G. & Gehrs, C.W. (1982). Ecosystem risk analysis: A new methodology. - Environ. Tox. & Chem., 1:167-177.

Persson, C., Rodhe, H. and De Geer, L.-E., 1987. The Chernobyl accident - A meteorological analysis of how radionuclides reached and were deposited in Sweden. - Ambio, 16:20-31.

Petersen Jr, R.C., Landner, L. and Blanck, H., 1986. Assessment of the impact of the Chernobyl reactor accident on the biota of Swedish streams and lakes. - Ambio, 15:327-331.

Pfaffenberger, R.C. and Patterson, J.H., 1987. Statistical methods. - Irwin, Illinois, 1246 p.

Råd and riktlinjer för kalkning av sjöar and vattendrag, Medd. från Fiskeristyrelsen, Nr. 1, 1982. Recommendations and guidelines for liming of lakes and waterways (in Swedish).

Rocheller, B., Liff, C., Campbell, W., Cassell, D., Church, R. and Nusz, R., 1989. Regional relationships between geomorphic/hydrologic parameters and surface water chemistry relative to acidic deposition. - J. Hydrol., 112:103-120.

Rosenberg, R. (ed.), 1986. Eutrophication in marine waters surrounding Sweden. - SNV Report 3054, Solna, 137 p.

Saether, O.A., 1979. Chironomid communities as water quality indicators. - Holarctic Ecol., 2:65-74.

Santchi, P.H., Bollhalder, S., Zingg, S., Luck, A. and Farrenkothen, K., 1990. The self-cleaning capacity of surface waters after radioactive fallout. Evidence from European waters after Chernobyl, 1986-1988. - Environ. Sci. Technol., 24:519-527.

SNV, 1980. Riktvärden för vattenkvalitet. Kadmium, förekomst, miljöeffekter, riktvärden för sjöar och vattendrag. - In: Wiederholm, T. (mimeo), Uppsala.

Wallin, M., Håkanson, L. and Persson, J., 1990. Eutrophication of coastal waters. Load models for nutrients (in Swedish with English summary). - Uppsala univ., Naturgeografiska inst., 224 p.

Wetzel, R.G., 1975. Limnology. - Saunders, Philadelphia, 743 p.

Vemuri, V., 1978. Modeling of complex systems. - Academic Press, 448 p.

Vollenweider, R.A., 1968. The scientific basis of lake eutrophication, with particular reference to phosphorus and nitrogen as eutrophication factors. - Tech. Rep. DAS/SCI/68.27, OECD, Paris, 159 p.

Vollenweider, R.A., 1976. Advances in defining critical loading levels for phosphorus in lake eutrophication. - Mem. 1st. Ital. Idrobiol., 33:53-83.

8. APPENDIX

1. Lake data from the 15 well-studied lakes and all 41 lakes.
2. Lake parameters and abbreviations.

Appendix 1A : Data from the 15 best studied lakes

EFFECT PARAMETERS (Bq/kg ww) · DOSE PARAMETERS

Lake	Cs-pi87	Cs-pi88	Cs-pe86	Cs-pe87	Cs-pe88	Cs-soil (Bq/m^2)	Cs-wa87 (Bq/l)	Cs-wa88	Cs-su86 (Bq/kg dw)	Cs-bo86	Cs-su87	Cs-bo87	Cs-su88	Cs-bo88
2110	322	323	2300	1502	451	3000	0,04	0,01	2800	8400	500	1060	1036	930
2117	942	2088	16900	10600	2857	10000	0,12	0,13	25400	36700	1870	670	1772	1921
2119	821	1359	12700	5500	2674	4000	0,13	0,1	9200	19200	2700	1600	813	1102
2120	260	791	5200	1754	586	4000	0,03	0,07	2700	3300	5600	4400	646	311
2121	2009	3909	25700	10618	8393	15000	0,14	0,14	31400	37700	1400	410	2776	3450
2122	1191	2821	27000	7951	3852	15000	0,16	0,17	25000	29800	860	800	2172	2486
2201	4654	7881	51800	33100	17588	70000	0,79	0,66	158000	576000	8467	18520	6682	6907
2206	1932	4382	23500	18100	6521	35000	0,51	0,28	36600	68100	4149	5312	3489	1859
2212	4147	6212	76200	41600	16335	50000	0,5	0,4	52100	241000	6763	10890	4961	5271
2213	4444	9687	96000	23400	8919	35000	0,64	0,43	97100	137000	15450	20973	10347	11534
2214	6540	7380	49510	30000	12839	70000	0,64	0,64	69000	115000	10077	15590	5855	7890
2215	1683	2423	23800	20900	3250	40000	0,15	0,09	41300	95100	6084	10878	4944	3582
2216	2692	3535	44794	24570	5259	50000	0,61	0,29	65300	56200	16953	23145	13404	10450
2217	769	•	6400	4986	1487	50000	0,18	0,07	20800	19100	7510	7330	6102	6012
2218	818	916	9100	2707	2196	40000	0,07	0,06	6500	6400	3874	5246	2838	2803

LAKE MORPHOMETRIC PARAMETERS · CATCHMENT PARAMETERS

Lake	Area (km^2)	Vol (km^3)	Dm (m)	DR	BA (%)	Q (m^3/s)	T (yr)	ADr (km^2)	Br	Lake% (%)	For%	Mire%	Ol%	Rock%	Mor%	Im%	Acid%
2110	0,44	0,002	4,7	0,14	86,4	0,13	0,48	13	34	1	86	12	0	6	81	5	95
2117	0,58	0,002	4,2	0,18	86,1	0,15	0,53	14	34	4	62	33	1	0	63	44	56
2119	0,15	0,00041	2,7	0,14	73,7	0,01	1,2	1	60	3	93	3	0	0	93	0	100
2120	0,16	0,00033	2,1	0,19	78,1	0,03	0,3	3	44	0	96	4	0	0	96	0	96
2121	0,12	0,00029	2,4	0,14	81,7	0,02	0,47	2	194	0	76	15	9	14	53	0	100
2122	0,16	0,00045	2,8	0,14	85	0,04	0,38	4	84	0	83	17	0	19	39	79	21
2201	0,25	0,002	8,3	0,06	35,3	0,2	0,33	20	56	1	89	10	0	20	63	91	5
2206	0,59	0,004	6,4	0,12	80,2	0,26	0,48	25	46	1	77	21	1	24	72	58	42
2212	0,28	0,001	4,9	0,11	78,4	0,18	0,24	18	53	5	79	16	0	24	50	91	1
2213	0,47	0,002	5,1	0,13	81,4	0,05	1,6	4	80	0	86	11	3	5	74	100	0
2214	0,07	0,0002	2,9	0,09	72,1	0,39	0,02	38	42	1	84	13	1	21	48	100	0
2215	0,24	0,001	4,5	0,11	78,4	0,08	0,44	8	80	0	83	9	7	12	31	14	86
2216	0,16	0,001	4	0,1	75,9	0,06	0,37	6	45	7	78	4	11	34	18	95	5
2217	0,18	0,001	3,8	0,11	78,4	0,11	0,21	12	62	4	73	3	20	20	24	93	7
2218	0,25	0,001	2,1	0,24	84,6	0,17	0,1	17	54	1	83	8	8	4	56	3	97

Appendix 1B : Data from the 15 best studied lakes

WATER CHEMICAL PARAMETERS 1986 TO 1988

Lake	temp-86 (degrees C)	pH-86	alk-86 (meq/l)	cond-86 (mS/m)	Col-86 (mg Pt/l)	Sec-86 (m)	totP-86 (µg/l)	Fe-86 (µg/l)	Ca-86 (meq/l)	CaMg-86 (meq/l)	K-86 (µeq/l)	temp-87 (grad C)	pH-87	alk-87 (meq/l)
2110	14,53	6,68	0,168	3,62	100	2,17	14,6	468	0,238	0,295	•	13,1	7,1	0,322
2117	14,26	5,78	0,018	2,12	88	1,905	12,6	534	0,107	0,168	•	14,8	6,72	0,15
2119	15,95	6	0,016	1,88	75	2,156	11,9	331	0,097	0,112	•	13	6,14	0,017
2120	12,5	6,36	0,061	3,48	105	2,078	9,6	527	0,212	0,25	•	14,5	6,88	0,257
2121	14,14	6,58	0,099	2,94	98	2,056	13,2	792	0,188	0,24	•	13,7	6,7	0,091
2122	15,78	5,84	0,035	2,14	99	2,171	11,6	553	0,103	0,134	•	13,9	6,66	0,132
2201	11,14	5,36	0,008	2,4	112	•	11,4	402	0,1	0,147	•	10,4	6,38	0,098
2206	13,7	5,74	0,024	2,02	111	•	13,4	711	0,106	0,142	•	11,6	5,85	0,03
2212	11,96	6,17	0,054	2,28	74	•	10,8	385	0,128	0,162	•	11,7	6,56	0,142
2213	14	6,67	0,076	2,96	39	•	8,6	150	0,146	0,214	•	11,6	6,6	0,08
2214	10,5	5,1	0,006	2,34	114	•	11,8	447	0,098	0,13	•	9,9	6	0,056
2215	12,48	6,57	0,084	2,84	66	•	12,2	217	0,142	0,192	•	11,1	6,75	0,138
2216	14,12	6,77	0,226	5,6	39	•	14,2	170	0,298	0,4	•	12,1	6,78	0,216
2217	12,46	6,93	0,298	7,4	48	•	27,8	382	0,404	0,532	•	12,2	6,92	0,288
2218	11,94	6,83	0,182	4	67	•	17,4	339	0,224	0,302	•	11,4	6,72	0,182

Lake	cond-87 (mS/m)	Col-87 (mg Pt/l)	Sec-87 (m)	totP-87 (µg/l)	Fe-87 (µg/l)	Ca-87 (meq/l)	CaMg-87 (meq/l)	K-87 (µekv/l)
2110	4,38	88	1,8	9	304	0,336	0,372	35
2117	3,7	111	1,8	8	500	0,282	0,274	15,333
2119	1,92	70	2,4	8,3	259	0,087	0,112	•
2120	4,875	90	1,9	14,1	410	0,393	0,367	•
2121	2,76	86	2,3	10,9	583	0,174	0,21	•
2122	3	96	1,9	8,9	520	0,214	0,222	•
2201	2,9	150	1,8	10,5	398	0,192	0,238	11,8
2206	1,98	145	1,8	13,3	597	0,084	0,118	7,1
2212	3,04	88	2,5	8	256	0,209	0,258	6,4
2213	3,08	53	3,6	8,7	187	0,132	0,206	11
2214	2,54	138	1,7	9,7	472	0,156	0,2	9,533
2215	3,36	82	2,4	8,9	231	0,196	0,252	12,05
2216	5,64	48	2,6	11,4	195	0,286	0,4	26,45
2217	7,34	71	1,7	18,3	311	0,376	0,52	30,5
2218	5,3	88	2,5	13,2	295	0,212	0,316	12,35

Lake	temp-88 degrees	pH-88	alk-88 (meq/l)	cond-88 (mS/m)	Col-88 (mg Pt/l)	Sec-88 (m)	totP-88 (µg/l)	Fe-88 (µg/l)	Ca-88 (meq/l)	CaMg-88 (meq/l)	K-88 (µeq/l)
2110	16,6	7,04	0,508	6,76	69	2,4	7,8	235	0,521	0,604	•
2117	18,7	6,78	0,17	3,46	93	2,5	7,66	323	0,242	0,28	•
2119	16,1	6,72	0,065	2,92	66	2,6	7,12	282	0,107	0,172	•
2120	20,1	6,66	0,168	4,44	89	1,8	14,66	500	0,281	0,344	•
2121	16,2	6,7	0,12	3,32	80	2,3	8,55	711	0,192	0,244	•
2122	15,5	6,78	0,17	3,58	82	2,2	6,12	519	0,247	0,286	•
2201	13,7	6,76	0,174	3,94	106	2,3	16	303	0,287	0,342	16,72
2206	13	6,56	0,132	3,2	139	1,8	13	549	0,222	0,276	18,14
2212	14,7	6,74	0,1	2,92	72	3	11,6	232	0,174	0,222	9,48
2213	15,2	6,8	0,118	3,72	49	3,6	7,8	191	0,158	0,268	12,5
2214	12,3	6,38	0,108	3,16	118	2,2	10,6	383	0,213	0,258	10,62
2215	15,3	6,68	0,14	3,56	70	3,1	8,94	226	0,188	0,252	9,18
2216	15,8	6,88	0,192	5,4	41	3,1	13,8	198	0,266	0,374	25,42
2217	16,1	7,04	0,334	7,94	52	2,4	23,8	319	0,395	0,564	28,78
2218	15,4	6,94	0,186	4,38	52	3,1	13,6	280	0,216	0,314	14,58

Appendix 1C: Data from the 41 lakes before remedial measures (March-86 to Feb.-87)

Lake	Cs-pe86	Cs-pe87	Cs-pe88	Cs-pi87	Cs-pi88	Cs-soil	pH	alk	cond	Colour	totP	Fe	Ca	CaMg	K
2101	12900	7295	1961	1158	1679	15000	6,2	0,04	4,3	145	11,3	878	0,18	0,25	•
2102	2800	514	1083	195	304	4000	6,3	0,08	3,5	135	26,4	1545	0,18	0,27	16,1
2103	9900	3289	2706	1358	1982	10000	6,1	0,05	3,2	95	10,9	1540	0,21	0,22	•
2104	11300	8654	2069	1132	1178	7500	6,1	0,04	2,6	131	8,8	1011	0,19	0,19	•
2105	7200	5593	812	531	728	4000	6	0,05	3,1	143	17,1	1200	0,23	0,24	•
2106	3390	1802	578	297	329	4000	6,2	0,06	3,7	140	10,3	728	0,23	0,27	•
2107	33400	19355	6574	3483	3746	15000	6,4	0,05	2,5	57	4,3	159	0,12	0,18	•
2108	5600	4232	1942	743	1088	4000	6,1	0,03	2,2	96	7,6	290	0,09	0,15	•
2109	8200	2500	415	417	612	4000	5,8	0,02	2,5	109	8,4	633	0,13	0,15	7
2110	2300	1502	451	322	323	3000	6,6	0,17	3,8	112	11,8	472	0,26	0,31	15
2111	1730	1236	1077	258	351	3000	6,4	0,08	2,7	83	6,4	290	0,17	0,18	•
2112	7160	5100	3412	623	1663	4000	5,6	0,02	2,2	189	11,7	1705	0,16	0,18	•
2113	13300	7544	2936	1784	3014	10000	5,7	0,03	2,9	201	14,2	1340	0,2	0,23	•
2114	2200	1688	1160	400	837	7500	6,2	0,08	4	153	13,3	1438	0,28	0,3	•
2115	4600	1826	739	285	616	4000	6,2	0,09	3,6	153	13	895	0,32	0,32	•
2116	4200	1637	359	537	844	2500	5,9	0,02	2	104	6,9	685	0,12	0,14	•
2117	16900	10600	2857	942	2088	10000	5,6	0,02	2,3	130	10,6	622	0,11	0,18	9
2118	15800	12480	9916	1346	2497	10000	6	0,04	2,4	110	8	1995	0,1	0,21	9
2119	12700	5500	2674	821	1359	4000	5,9	0,02	2	110	8,9	334	0,1	0,12	9
2120	5200	1754	586	260	791	4000	6,2	0,06	3,6	133	8,5	577	0,21	0,28	7,1
2121	25700	10618	8393	2009	3909	15000	6,4	0,11	3,1	115	11,6	1008	0,21	0,26	7,5
2122	27000	7951	3852	1191	2821	15000	5,8	0,03	2,2	110	10	651	0,11	0,16	7,5
2201	51800	33100	17588	4654	7881	70000	5,2	0,01	2,4	122	9,5	439	0,11	0,14	8,7
2202	81200	83300	22139	5797	8344	40000	5,5	0,04	3	42	7,7	163	0,14	0,17	•
2203	66616	42800	14274	5476	8863	50000	5,9	0,03	2,3	36	5	59	0,11	0,13	9,7
2204	32768	17700	3426	3016	4116	35000	6,2	0,08	2,8	69	8,3	244	0,17	0,21	8,6
2205	17577	14600	4400	1355	3217	35000	6,5	0,11	3,9	56	12	257	0,23	0,28	•
2206	23500	18100	6521	1932	4382	50000	5,6	0,03	2,1	128	11,9	770	0,1	0,14	8,3
2207	54334	97900	38206	6813	13756	50000	5,4	0,01	1,6	65	6,3	258	0,08	0,1	5,9
2208	74299	55000	18470	5932	9164	50000	5,9	0,04	2,4	73	6,8	332	0,11	0,16	8,8
2209	37400	23000	4611	2194	3335	40000	6,2	0,07	2,6	95	11,3	547	0,15	0,19	8,1
2210	•	57700	34731	7310	14721	50000	5,2	0,01	1,9	121	13,5	510	0,09	0,1	•
2211	524045	77724	31302	5919	11473	50000	6,1	0,05	2	58	7	318	0,12	0,13	•
2212	76200	41600	16335	4147	6212	50000	6	0,06	2,4	79	8,7	405	0,14	0,18	6
2213	96000	23400	8919	4444	9687	35000	6,4	0,08	3	38	7,1	166	0,15	0,22	8,4
2214	49510	30000	12839	6540	7380	70000	5,1	0,01	2,3	115	10,1	499	0,11	0,14	6
2215	23800	20900	3250	1683	2423	40000	6,2	0,08	2,8	70	9,5	269	0,13	0,19	10,5
2216	44794	24570	5259	2692	3535	50000	6,5	0,22	5,5	35	11,1	160	0,31	0,42	26,9
2217	6400	4986	1487	769	•	50000	6,6	0,37	8,2	46	22,9	356	0,48	0,64	36,7
2218	9100	2707	2196	818	916	40000	6,5	0,25	4,6	69	13,8	459	0,26	0,37	16,3
2219	58030	34100	23856	6483	9102	35000	5,9	0,03	1,9	109	7	430	0,13	0,14	•

Appendix 1D: Data from the 41 lakes before remedial measures (March-86 to Feb.-87)

Lake	Area	Vol	Dm	DR	BA	Q	T	ADr	Br	Lake%	For%	Mire%	Ol%	Rock%	Mor%	Im%	Acid%
2101	0,8	0,004	4	0,24	77,9	0,19	0,89	18	37	0	68	16	16	0	64	30	70
2102	0,35	0,00044	1,1	0,54	60	0,79	0,02	80	20	5	87	6	2	2	76	0	97
2103	0,73	0,002	2,3	0,36	49,3	0,04	1,19	4	14	0	78	13	9	1	85	50	42
2104	1,68	0,005	3,1	0,42	82	0,48	0,35	47	22	3	79	16	2	0	63	13	84
2105	0,15	0,00039	2,7	0,14	75	0,02	0,7	2	131	9	68	23	0	0	68	0	100
2106	0,11	0,0005	4,7	0,07	86,4	0,04	0,42	4	38	0	100	0	0	0	90	0	50
2107	1,31	0,006	4,3	0,27	84,4	0,1	1,74	9	73	1	79	14	5	9	60	20	80
2108	0,82	0,003	3,3	0,25	69,5	0,09	0,94	8	34	1	78	20	0	13	79	0	89
2109	0,22	0,001	3,2	0,15	63,6	0,16	0,14	16	44	1	91	7	0	5	68	1	98
2110	0,44	0,002	4,7	0,14	86,4	0,13	0,48	13	34	1	86	12	0	6	81	5	95
2111	0,4	0,001	2,1	0,32	83,3	0,3	0,09	35	44	10	77	12	0	1	76	6	71
2112	0,23	0,00038	1,7	0,28	69,6	0,03	0,46	2	36	1	70	27	1	0	72	100	0
2113	0,54	0,001	1,2	0,61	14,8	0,17	0,12	17	9	6	67	27	6	0	55	62	38
2114	0,35	0,00041	1,2	0,49	54,3	0,39	0,03	39	26	5	71	18	0	0	77	15	82
2115	0,13	0,00026	2	0,18	69,2	0,03	0,26	3	25	0	95	5	6	0	83	0	100
2116	0,93	0,003	3,5	0,28	69,5	0,31	0,34	31	25	5	65	28	0	0	67	0	95
2117	0,58	0,002	4,2	0,18	86,1	0,15	0,53	14	34	4	62	33	2	0	63	44	56
2118	0,38	0,0004	1,1	0,56	52,6	0,2	0,06	20	33	8	74	18	1	2	71	38	62
2119	0,15	0,00041	2,7	0,14	73,7	0,01	1,2	1	60	3	93	3	0	0	93	0	100
2120	0,16	0,00033	2,1	0,19	78,1	0,03	0,3	3	44	0	96	4	0	0	96	0	96
2121	0,12	0,00029	2,4	0,14	81,7	0,02	0,47	2	194	0	76	15	9	14	53	0	100
2122	0,16	0,00045	2,8	0,14	85	0,04	0,38	4	84	0	83	17	0	19	39	79	21
2201	0,25	0,002	8,3	0,06	35,3	0,2	0,33	20	56	1	89	10	0	20	63	91	5
2202	0,4	0,003	6,2	0,1	75,9	0,04	2,2	4	24	4	75	21	0	40	3	88	12
2203	1,7	0,017	10,1	0,13	81,4	0,19	2,9	17	42	4	84	12	0	13	64	77	23
2204	1,26	0,011	8,5	0,13	81,4	0,53	0,65	51	29	2	82	13	4	14	57	86	14
2205	1,05	0,007	6,9	0,15	83	0,21	1,1	23	46	1	91	4	4	24	51	96	0
2206	0,59	0,004	6,4	0,12	80,2	0,26	0,48	25	46	1	77	21	4	1	72	58	42
2207	2,02	0,01	4,9	0,29	84,2	0,13	2,4	10	52	0	64	35	1	2	62	85	15
2208	1,32	0,008	6	0,19	84,4	0,24	1,1	21	22	2	79	17	0	5	63	87	13
2209	0,63	0,005	8,1	0,1	75,3	0,19	0,85	19	35	2	79	19	2	9	46	0	100
2210	1,57	0,01	6,2	0,2	84,5	0,47	0,68	41	16	3	67	30	2	5	62	79	21
2211	2,7	0,016	6,1	0,27	84,4	0,33	1,6	27	23	3	67	28	2	4	63	62	38
2212	0,28	0,001	4,9	0,11	78,4	0,18	0,24	18	53	5	79	16	0	24	50	91	1
2213	0,47	0,002	5,1	0,13	81,4	0,05	1,6	4	80	1	86	11	3	5	74	100	0
2214	0,07	0,0002	2,9	0,09	72,1	0,39	0,02	38	42	0	84	13	1	21	48	100	0
2215	0,24	0,001	4,5	0,11	78,4	0,08	0,44	8	80	0	83	9	1	12	31	14	86
2216	0,16	0,001	4	0,1	75,9	0,06	0,37	6	45	7	78	4	7	34	18	95	5
2217	0,18	0,001	3,8	0,11	78,4	0,11	0,21	12	62	4	73	3	11	20	24	93	7
2218	0,25	0,001	2,1	0,24	84,6	0,17	0,1	17	54	4	83	8	20	4	56	3	97
2219	2	0,009	4,6	0,31	83,9	0,27	1,1	27	20	1	66	32	0	0	66	50	50

Appendix 2. Lake parameters and abbreviations.

A. Lake morphometric parameters	Abbreviation
Max. length (km)	Lmax
Max. depth (m)	Dmax
Shoreline length (km)	lo
Total area (km^2)	Atot
Water surface (km^2)	a (or area)
Lake volume (km^3)	V (or Vol)
Mean width (km)	Bm
Mean depth (m)	Dm
Shore irregularity (dimensionless)	F
Volume development (dimensionless)	Vd
Dynamic ratio (dimensionless)	DR
Areas of erosion & transportation (% of a)	BET
Areas of accumulation (% of a)	BA
Water discharge (m^3/s)	Q
Water retention time (yr)	T

B. Drainage area parameters	
Drainage area (km^2)	ADr
Relief near area (m/km^2)	NBr
Relief (m/km^2)	Br
Lake % (% of ADr)	Lake%
Forest%	For%
Mire%	Mire%
Open land%	Ol%
Rock%	Rock%
Moraine%	Mo%
Coarse sediment%	Coar%
Fine sediment%	Fine%
Basic rocks%	Bas%
Intermediate rocks%	Im%
Acid rocks%	Acid%

C. Water chemical parameters	
pH	pH
Alkalinity (meq/l)	alk
Conductivity (mS/m)	cond
Colour (mg Pt/l)	Col
Total phosphorus (µg/l)	totP
Iron (µg/l)	Fe
Calcium (meq/l)	Ca
Hardness (meq/l)	CaMg
Potassium (µeq/l)	K
Temperature (degree C)	Temp (or Tr)
Secchi depth (m)	Sec

D. Caesium parameters	
Cs-137 in soil (=fallout), (Bq/m2)	Cs-soil
Cs-137 in lake water (Bq/l)	Cs-wa
Cs-137 in 1-kg pike (Bq/kg ww)	Cs-pi
(Cs-pi87 means Cs-pi from 1987, etc)	
Cs-137 in 1+perch (Bq/kg ww)	Cs-pe
Cs-137 in surface sediment traps (Bq/kg dw)	Cs-su
Cs-137 in bottom sediment traps (Bq/kg dw)	Cs-bo